ULTRACONDENSED MATTER BY DYNAMIC COMPRESSION

Dynamic compression is an experimental technique with interdisciplinary uses, ranging from enabling the creation of ultracondensed matter under previously impossible conditions to understanding the likely cause of unusual planetary magnetic fields. Readers can now gain an intuitive understanding of dynamic compression; clear and authoritative chapters examine its history and experimental method, as well as key topics including dynamic compression of liquid hydrogen, rare gas fluids and shock-induced opacity. Through an up-to-date history of dynamic compression research Nellis also clearly shows how dynamic compression addresses and will continue to address major unanswered questions across the scientific disciplines. The past and future role of dynamic compression in studying and making materials at extreme conditions of pressure, density and temperature is made clear, and the means of doing so are explained in practical language perfectly suited for researchers and graduate students alike.

WILLIAM J. NELLIS is an Associate of the Department of Physics at Harvard University, a Fellow of the American Physical Society, Division of Condensed Matter Physics, holder of the APS Duvall Award for Shock Compression Science, former Chair of the APS Topical Group on Shock Compression of Condensed Matter, former President of the International Association of High Pressure Science and Technology and holder of its Bridgman Award. He has performed extensive dynamic compression research at Lawrence Livermore National Laboratory and published more than 250 papers in various scientific journals.

ULTRACONDENSED MATTER BY DYNAMIC COMPRESSION

WILLIAM J. NELLIS
Harvard University, Massachusetts

CAMBRIDGE
UNIVERSITY PRESS

University Printing House, Cambridge CB2 8BS, United Kingdom

One Liberty Plaza, 20th Floor, New York, NY 10006, USA

477 Williamstown Road, Port Melbourne, VIC 3207, Australia

4843/24, 2nd Floor, Ansari Road, Daryaganj, Delhi – 110002, India

79 Anson Road, #06–04/06, Singapore 079906

Cambridge University Press is part of the University of Cambridge.

It furthers the University's mission by disseminating knowledge in the pursuit of education, learning, and research at the highest international levels of excellence.

www.cambridge.org
Information on this title: www.cambridge.org/9780521519175
DOI: 10.1017/9781139031981

First published 2017

Printed in the United States of America by Sheridan Books, Inc.

A catalogue record for this publication is available from the British Library.

Library of Congress Cataloging-in-Publication Data
Names: Nellis, W. J.
Title: Ultracondensed matter by dynamic compression / William J. Nellis, Harvard University, Massachusetts.
Description: Cambridge : Cambridge University Press, 2017. | Includes bibliographical references and index.
Identifiers: LCCN 2016054367 | ISBN 9780521519175 (Hardback : alk. paper)
Subjects: LCSH: Condensed matter. | Materials at high pressures. | High pressure (Science) | High pressure geosciences.
Classification: LCC QC173.454 .N45 2017 | DDC 530.4/1–dc23
LC record available at https://lccn.loc.gov/2016054367

ISBN 978-0-521-51917-5 Hardback

Contents

Preface

The science of dynamic compression began in 1870 when W. J. M. Rankine published a paper on the conservation equations of momentum, mass and energy across the front of a supersonic shock wave in an ideal gas. The paper was published in *Philosophical Transactions of the Royal Society of London.* Rankine was a professor at the University of Glasgow and a colleague of William Thomson, Lord Kelvin, also of the University of Glasgow.

Supersonic hydrodynamics was developed mathematically in Western Europe in the last half of the nineteenth century. H. Hugoniot derived Rankine's conservation equations on a more general basis in the 1880s. With the advent of quantum mechanics in the 1920s, dynamic compression would probably have been completely forgotten but for the threat of World War II in the 1930s. In 1940, H. Bethe and E. Teller wrote the first theoretical treatment of thermal equilibration in the front of a shock wave in an ideal gas.

World War II generated substantial governmental funding for experimental facilities in both the United States and the Soviet Union, which enabled tests of previous theoretical predictions about shock wave propagation. A major emphasis of that period was development of fast experimental techniques to measure pressure-volume data under shock compression using the Rankine-Hugoniot shock wave conservation equations.

In the 1950s, researchers using dynamic and static compression combined to determine pressure, density and likely crystal structure of the α-ε transition of Fe at 13 GPa. That determination was the first generally accepted phase transition observed under shock compression and with that acceptance shock compression was recognized generally as a science by the static high pressure community. It was also the first fixed-point pressure standard derived with dynamic compression.

In subsequent years, dynamic compression experiments were performed primarily in defense laboratories and in a few universities and companies. As a result, few textbooks have been written on dynamic compression, although several have been written on shock compression, which is a particular type of dynamic compression. Because of this, researchers have often had to learn about dynamic compression from a relatively few published papers and a few unpublished reports. Thus research in dynamic compression had become essentially isolated from the scientific community.

Then it happened. In 1996, metallic fluid H (MFH) was made under dynamic compression at finite temperatures in a crossover from semiconducting H to poor metallic (degenerate) H with measured Mott's minimum metallic conductivity. This crossover completes at the density of the insulator-metal transition from solid H_2 to solid H predicted by Wigner and Huntington in 1935. Finding MFH experimentally gave dynamic compression visibility it never had, but there also suddenly arose the need to explain what it is exactly. There was, however, no easy way to explain it – no book to which to refer people.

So I have written this book to be of general interest to undergraduate and graduate students, for professors that teach them and for research scientists at national laboratories and in industry who need to know it. This book is not intended to be an all-inclusive review. It is about ideas and concentrates on pressures greater than ~10 GPa, below which traditional publications cover shock compression. I am trying to convey the idea that dynamic compression at much higher dynamic pressures is a vehicle for novel scientific research that has lead to exciting scientific discoveries. To this end this book contains a chapter on the basics of dynamic compression needed to design and understand results of such experiments and a few examples that illustrate how this technique connects to understanding general scientific questions that have been unresolved for years.

Implied by these discussions is the fact that a new regime of thermodynamic conditions has been opened up for experimental investigations and associated theory. Dynamic compression discussed herein is an experimental technique, not an academic discipline. Opportunities are available in physics, chemistry, planetary science, materials science and other fields. Uses of dynamic compression are limited only by the imagination of its practitioners. In this regard I am reminded of the words of William Fowler, former President of the American Physical Society: "We look forward to the future of our profession as an intellectual enterprise for the individual and as a practical enterprise for society."

Acknowledgments

This research was performed in collaboration with colleagues at Lawrence Livermore National Laboratory and at laboratories and universities in the international high pressure community in Japan, Russia, China and Sweden. In particular I would like to acknowledge A. C. Mitchell, M. Ross, N. C. Holmes, R. Chau, N. W. Ashcroft, F. H. Ree, I. F. Silvera, T. Mashimo, G. I. Kanel, X. Zhou, N. Ozaki and R. Ahuja.

1
Introduction

Dynamic compression is a nonlinear process that achieves extreme pressure P, density ρ, internal energy E and temperature T via a supersonic compressional wave. Because compression is so fast, temperature and entropy S are generated simultaneously under adiabatic compression within the front of the supersonic matter wave. Thermodynamic states achieved are tunable over wide limits by choice of the temporal history of the pressure pulse that generates the compression. Magnitudes of thermodynamic states are determined by the magnitude of energy coupled into a material in the process.

Pressure is defined herein to be dynamic if its rate of application affects induced T and S and the compression is adiabatic. Supersonic adiabatic compression significantly affects T and S. In contrast "slow" sonic (isentropic) compression is reversible by definition and its rate of application has relatively small affect on T. Static compression is sufficiently slow that heat is transported out of a sample essentially as it is produced by compression and thus is isothermal and non-adiabatic.

Supersonic hydrodynamic shock flow in a fluid generates a matter wave, whose front is a jump in pressure and associated thermodynamic variables from a lower-pressure state ahead of the front of the wave to a higher-pressure state behind the wave front. Because shock flow is supersonic, material ahead of the shock wave in a fluid does not receive a precursor signal prior to arrival of that wave. Thus, a supersonic wave "snow plows" a fluid, which rapidly compresses, heats and disorders matter upon thermally equilibrating to final pressure behind the wave front. The front of a shock wave in a fluid is the non-equilibrium region in which a shocked fluid equilibrates thermally upon going from the lower-pressure to the higher-pressure states.

Historically, dynamic compression (DC) has been achieved primarily with a single, sharp, step-increase in pressure, which in a condensed fluid has negligible rise time (~ps) over a thin (~nm) wave front. Such a wave is commonly known as a shock wave in condensed matter. Dynamic compression is not restricted to a Heaviside-type step-jump of shock pressure, as is often thought. Probably the most important advance in dynamic compression in recent years is recognition of the fact that the pulse driving the compression is not restricted to a shock wave. This recognition has expanded significantly the range of thermodynamic states achieved dynamically, and with it the capability to address an expanded number of challenging problems over a wide range of pressures and temperatures.

A paradigm in this regard is the making of metallic fluid hydrogen (MFH), which has been a major goal of high-pressure experiments since the 1950s. MFH has been made by shaping applied dynamic pressure pulse to obtain quasi-isentropic compression with a multiple-shock wave consisting of ~10 small shocks to peak pressure. Such a wave achieves substantially lower temperatures, entropies and higher densities than compression with a single sharp shock. This technique made MFH degenerate at pressure 140 GPa (1.4 million bar) and 0.63 mol H/cm^3 (ninefold liquid-hydrogen density) (Weir et al., 1996a; Nellis et al., 1999), which is essentially the metallization density of H predicted by Wigner and Huntington (1935).

That metallization of monatomic fluid H was achieved by overlap of 1s^1 electronic wave functions on adjacent H atoms in experiments with lifetimes of ~100 nanoseconds (ns) (Weir et al., 1996a; Pfaffenzeller and Hohl, 1997; Nellis, et al., 1998; Nellis et al., 1999; Fortov et al., 2003). The making of MFH under dynamic compression illustrates the importance of temperature for achieving a metallic fluid at a relatively low pressure and the broad potential of DC for addressing major long-standing scientific issues.

Bethe and Teller (1940) performed the first theoretical investigation of the nature of the front of a shock wave. That work of Bethe and Teller (BT) showed that the asymptotic values of material velocity, density, pressure and temperature at high pressures a sufficient distance from the front of the shock wave (Fig. 1.1) are determined by the values of those quantities at low pressures ahead of the traveling shock front, independent of microscopic non-equilibrium conditions within the shock front. That unpublished report of BT is considered to be a key paper in the study of solids far from equilibrium (Mermin and Ashcroft, 2006).

Spatial and temporal thicknesses of a 1.2 GPa shock front in compressible liquid Ar were calculated by Hoover (1979) and by Klimenko and Dremin (1979) using Navier-Stokes equations and molecular dynamics, respectively. As shown in Fig. 1.1, spatial and temporal thicknesses of that shock front are ~nm and ~ps, respectively, which are sufficiently fast to generate substantial temperature T and thermally equilibrated disorder S. Multiple-shock compression is a sequence of several discreet jumps each as in Fig. 1.1. Thermally equilibrated heat and disorder in the form of temperature T and entropy S distinguish fast (~ps) shock compression from slow (~s) isothermal static compression. The apparently smooth increase

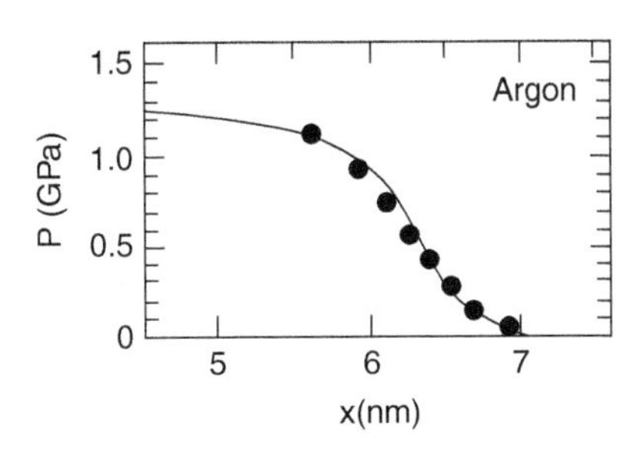

Fig. 1.1. Calculated rise to shock pressure of 1.2 GPa in shock front of liquid Ar. Full curve calculated with atomistic molecular dynamics (Klimenko and Dremin, 1979); solid circles calculated with Navier-Stokes equations (Hoover, 1979). Shock speed u_s is 1.8 km/s; rise time from 0 to 1.2 GPa is ~10^{-12} s over ~nm shock front thickness. Copyright 1979 by American Physical Society.

in pressure in Fig. 1.1 belies the fact that *P* and other calculated thermodynamic quantities were obtained by averaging computational results at various depths within the non-equilibrium shock front.

In the 1970s and 1980s, shock research was broadened to achieve quasi-isentropic compression by increasing the rise time of a shock wave. In planar geometry, ramp waves were generated experimentally with graded-density impactors (Barker and Hollenbach, 1974; Barker, 1984). A ramp wave is a pressure pulse that increases with time, generally linearly in practice. Multiple-shock waves were generated computationally using multiple layers of materials to break up a strong shock into numerous smaller ones (Lyzenga and Ahrens, 1982). Dynamic compression is achieved by a general pressure pulse shape, of which shock, multiple-shock and ramp-compression waves are three types. Because the rate of application of dynamic pressure *P* can readily be tuned experimentally, *T*, *S*, ρ and *E* are tunable over significant ranges by choice of applied pressure pulse *P(t)*, where *t* is time.

Dynamic compression is a mature field of hydrodynamics with its mathematical origins in European universities in the last half of the nineteenth century (Courant and Friedrichs, 1948). In the nineteenth century, what is known today as dynamic compression was probably known as supersonic hydrodynamics. The existence of dynamic compression might well have been forgotten with development of quantum mechanics in the 1920s, but for defense research initiated in the 1930s prior to World War II (Bethe and Teller, 1940). In the 1940s, funding appeared for national facilities to perform experimental shock-compression research to complement theoretical research performed between ~1850 and ~1910 (Chapter 4). Shock-compression research eventually morphed into dynamic compression. The purpose of this book is to provide a primer on the basic ideas of dynamic compression and to discuss recent achievements that illustrate its broad potential.

1.1 Beyond Shock Compression: Tunable Thermodynamics

In 1935, Wigner and Huntington (WH) predicted that at some high pressure greater than 25 GPa (0.25 Mbar), a density of 0.62 mol H/cm^3 and very low temperatures, electrically insulating solid H_2 would undergo a first-order transition to metallic H. Stewart (1956) made solid H_2 at the then "high" static pressure of 2 GPa, which generated interest in the static-pressure community to look for solid metallic H at much higher pressures. In the 1990s, MFH was made by specifically tuning thermodynamic conditions in liquid hydrogen by dynamic compression to dissociate liquid H_2 into fluid H at sufficiently high densities and low temperatures to make electrically conducting monatomic fluid H at sufficiently low temperatures to make degenerate fluid H.

Thermodynamics are tuned by tuning supersonic hydrodynamics. The shape of a dynamic pressure pulse and associated shock-induced dissipation were used to tune states off a shock-compression curve, which is commonly known as a Hugoniot or Rankine-Hugoniot curve. A shock-compression curve is a locus of states achieved adiabatically with a sequence of single, sharp shock compressions of increasing shock pressures. Because dynamic temperatures T are relatively high herein, S produced by dynamic compression is usually considered to be thermally equilibrated disorder. Because Helmholtz free energy F is given by $F = E - TS$, dissipation energy $E_d = TS$ is used to tune phase stability. In a given process S is maximized to minimize H at a given T.

In the 1990s, the H_2 vibron (diatomic vibrational frequency) was observed in the insulating solid at static pressures up to ~300 GPa. The H_2 molecule in the solid is extremely stable under pressure. WH had anticipated the possibility that their predicted simple dissociative transition might not occur and speculated on what else might happen instead at some very high pressure (Wigner and Huntington, 1935). By the early 1990s, metallic solid H had yet to be made under static compression, and dynamic compression was beginning to look attractive as a potential alternative tool to make metallic hydrogen by tuning dissipation S and T. The challenge was to find an appropriate pressure profile that would generate sufficient thermal energy and entropy via dissociation of H_2 to H at sufficiently high ρ and sufficiently low T to make degenerate metallic H.

The discipline of dynamic compression began when W. J. M. Rankine (1870), a professor at the University of Glasgow, published conservation equations of momentum, mass and internal energy across the front of a shock wave. Rankine explicitly used the equation of state (EOS) of an ideal gas, whose derivation had then been completed relatively recently in the sense that W. Thomson, Lord Kelvin, also of the University of Glasgow, had derived absolute 0 K on the Centigrade temperature scale (Thomson, 1848). H. Hugoniot was captain at the

Marine Artillery Academy of France and subsequently derived those conservation equations on a more general basis (Hugoniot, 1887, 1889; Cheret, 1992).

Conservation equations for *P*, *V* and *E* across the front of a shock wave are called the Rankine-Hugoniot (R-H) equations or simply the Hugoniot equations. The locus of states measured under shock compression is commonly called a Rankine-Hugoniot (R-H) curve, a Hugoniot or a shock adiabat. Experiments to measure values of (P_H, V_H and E_H) are called Hugoniot experiments, where P_H, V_H and E_H are values of *P*, *V* and *E* on the R-H curve. Dynamic compression in general is a subfield of Applied Mathematics (Liu, 1986).

A dynamic isentrope is the limit of an infinite number of infinitely weak adiabatic shock compressions (Zeldovich and Raizer, 1966). Dynamic quasi-isentropic (Q-I) compression is essentially an isentrop plus sufficient *S* and *T* to achieve significantly higher densities and lower temperatures than achievable with shock compression, and thus might drive dissociation of H_2 and induce metallization of H. Multiple-shock compression is quasi-isentropic – typically an initial relatively weak shock followed by sufficiently many relatively weak shocks that lie on the isentrope starting from the first shock state, as for an ideal gas (Nellis, 2006a).

Ramp compression is achieved by a continuous adiabatic increase in dynamic pressure, rather than a sequence of relatively small discreet step-increases in pressure, as in multiple-shock compression. Dissipation in ramp compression can be tuned by the magnitude of the initial shock and temporal slope of the ramp. The smaller the initial shock and the slower the increase with time of the ramp, the lower the final temperature achieved. Multiple-shock and ramp compressions are essentially equivalent in the sense they can be tuned to obtain similar states.

Fig. 1.2 shows pulses achieved by a single shock to a given pressure and by multiple shocks to the same pressure. Fig. 1.2 illustrates that increasing the number of shocks to a given pressure is equivalent to slowing down the rise time to that given pressure. Fig. 1.3 illustrates schematically the two curves in *P*-ρ space that correspond to the pressure histories in Fig. 1.2, plus the 0-K isotherm. P_H in

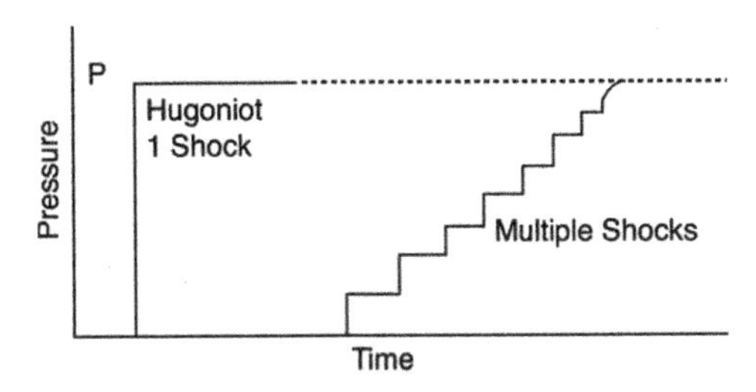

Fig. 1.2. Schematic of temporal pressure profiles for single- and multiple-shock compression to same pressure.

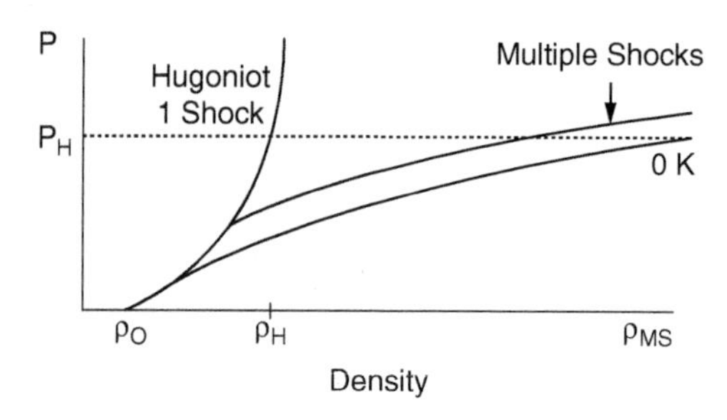

Fig. 1.3. Schematic in pressure-density space of single-shock (Hugoniot), multiple-shock (quasi-isentrope) and 0-K isotherm (static compression). P_H is P on Hugoniot at density ρ_H, which has limiting (maximum) shock density for given Hugoniot. ρ_{MS} is density achieved on multiple-shock compression to P_H. $\rho_{MS} > \rho_H$.

Fig. 1.3 is P on the Hugoniot at density ρ_H, which has a limiting (maximum) shock density for a given Hugoniot. ρ_{MS} is density achieved on a multiple-shock compression curve to pressure P_H. Fig. 1.3 shows $\rho_{MS} > \rho_H$. Extending the temporal interval of applied pressure history (Fig. 1.2) increases densities and decreases temperatures of thermodynamic states achieved by multiple-shock relative to single-shock compression.

If a material has internal degrees of freedom that can absorb shock-induced internal energy, then still lower Ts might be expected in addition to what would be produced hydrodynamically by multiple-shock compression alone. In the case of liquid H_2, dissipation energy is absorbed internally by dissociation of H_2 to H, producing entropy. This idea was successfully tested experimentally by producing metallic fluid H (Weir et al., 1996a; Nellis et al., 1999).

Ramp compression is a simple variation of multiple-shock compression, as illustrated in Fig. 1.4. A ramp-compression wave replaces discrete, numerous, weak, multiple shocks, as in Fig. 1.2, with a pressure that increases continuously from an initial weak shock pressure, say P_{H1}, up to maximum pressure P_{max}. Ramp compression is often a linear increase of pressure P with time. If $P_{H1} = 0$ and the ramp increase in P is sufficiently slow, then that ramp compression produces a dynamic isentrope. If $P_{H1} << P_{max}$ and the ramp increase in P is sufficiently slow, then that ramp compression is said to produce a quasi-isentrope.

1.2 Cold, Warm and Hot Matter

Ultracondensed matter herein is matter compressed by a factor ranging from ~1.5 to as much as 15-fold of initial sample density ρ_0 by pressures from ~50 GPa (0.05 TPa = 0.5 Mbar) to as much as a few TPa at relatively low temperatures T such that $T/T_F<<1$, where T_F is Fermi temperature of a Fermi-Dirac electron

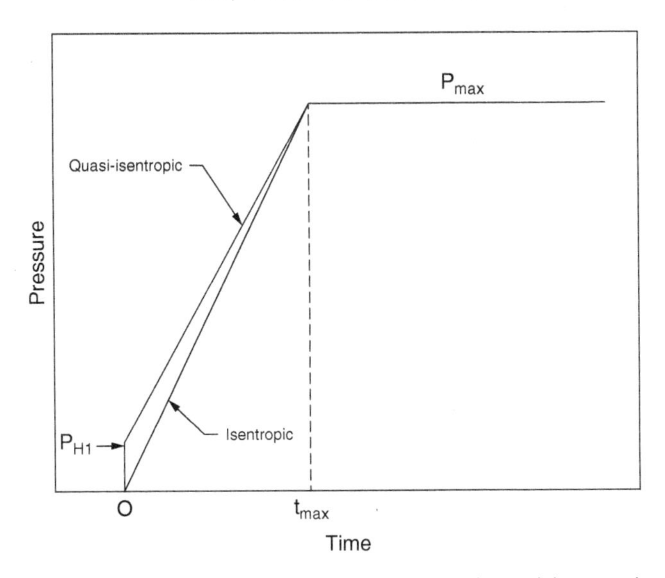

Fig. 1.4. Pressure histories for dynamic isentropic and quasi-isentropic compressions. t_{max} typically ranges from a few ns to ~100 ns (Nellis, 2006a).

system (Mott, 1936) and T/T_F is quantum degeneracy factor. T_F is defined at 0 K and depends on itinerant electron density. Absolute temperatures considered herein are typically in the range from 1000 K to tens of thousands of degrees Kelvin. Because ultracondensed compressions are large, T_F is also large. For this reason, ultracondensed degenerate matter is relatively straightforward to obtain, particularly for compressible materials. For example, MFH is made by dynamic compression at ~3000 K, ninefold atomic H density in liquid H_2 initially at 20 K and $T_F \approx 220{,}000$ K. Assuming total dissociation at that density, $T/T_F \approx 0.014 << 1$. MFH is "cold" at 3000 K because its metallization density ρ_{met} is so large and thus so is T_F.

While our chief interest is ultracondensed matter, dynamic compression can achieve high temperatures and degeneracy factors as well. In general, $T/T_F << 1$, $T/T_F \approx 1$ and $T/T_F >> 1$ for cold condensed matter, warm dense matter (WDM) and hot plasmas, respectively. Thus, while dynamic compression is a process of classical physics, states produced by dynamic compression span the quantum range from degenerate to highly non-degenerate.

A remarkable unexpected case of systematic behavior of warm dense fluid metals has been observed under shock pressures from 0.3 TPa up to 20 TPa (200 Mbar). Those extreme pressures were generated in proximity to underground nuclear explosions (Trunin, 1998) and discussed by Ozaki et al. (2016) and in Chapter 10. Estimated temperatures in those experiments range up to several hundred thousand degrees Kelvin – surprisingly large for such systematic behavior.

1.3 Experimental Timescales

Dynamic compression discussed herein is such that (1) the compression process is adiabatic and (2) the rate of application of pressure is sufficiently large that it affects temperatures and entropies achieved. That is, conditions achieved depend not only on the amount of energy deposited but also on the rate at which energy is deposited. Dynamic compression is adiabatic if insufficient time is available during experimental lifetime for heat transport at sonic speeds into or out of a sample under supersonic dynamic compression. This condition depends on wave speeds and sample dimensions.

For this reason, dynamic-compression experiments to measure material properties are generally performed in one-dimensional (1D) geometry, which enables thermodynamic conditions to be determined simply by supersonic hydrodynamics, independent of thermal transport. In this case thermodynamic states to determine physical properties are sensitive primarily to dynamic compression and pressure release in the direction parallel to supersonic wave propagation, commonly called the longitudinal direction. Simultaneously, pressure release and thermal losses at sonic velocities in the direction normal to the direction of supersonic compression can be virtually eliminated in a portion of sample volume. In this way, for example, a sample volume can be designed to have constant, uniform P, V, and T for a time called the experimental lifetime. To obtain such a state with a lifetime sufficiently long to make an accurate measurement with a given diagnostic resolution time, dynamic experiments must be designed for particular sample materials, sample holder geometries and dynamic waveforms.

Experimental timescales and sample thicknesses are determined by transit times of supersonic and sonic waves. An estimate of representative experimental lifetimes is obtained from the transit time across ~mm, which for Cu at sonic velocity, for example, means a transit time of ~200 ns (ns = 10^{-9} s). Transit times at supersonic velocities are smaller. Experimental lifetimes of dynamic experiments generally range from a few ns to a few hundred ns or more, depending on shock driver and pressure. Samples thicknesses range from a few μm (10^{-6} m) to ~few mm with sample surfaces made and inspected flat and parallel to ~μm or less depending on shock driver. Time resolution required for accurate measurements extends down to as small as ~10 ps (ps = 10^{-12} s) for fast laser-driven shock waves. Diagnostics with high time and spatial resolution, including tight control on synchronization and cross timing between triggering and diagnostic systems, are required for dynamic compression experiments. *Because of this fast timescale, thermal diffusion, mass diffusion, fluid turbulence and convection are generally negligible in experimental volumes diagnosed under adiabatic dynamic compression.*

1.4 Thermal Equilibrium

Compressible fluids and metals are generally in thermal equilibrium under dynamic compression. Atoms and molecules thermally equilibrate after sufficient energy is exchanged in a sufficiently large number of interatomic collisions. This process in fluids occurs within the front of a shock wave (Fig. 1.1), which produces thermally equilibrated matter in a time much less than the duration of the experiment.

To diagnose a thermally equilibrated state, a large number of inter-atomic collisions must occur within the resolution time of a diagnostic system. If diagnostic time resolution corresponds to more than ~10 collision times, the sample is probably in thermal equilibrium. Collision times are estimated by average inter-atomic distance divided by particle velocity, often assumed to be ideal-gas thermal velocity for estimation purposes. For MFH at 3000 K and ninefold H density, ns time resolution corresponds to ~10^4 proton-proton collisions, which means MFH is in thermal equilibrium. In contrast, if diagnostic time resolution is less than required to achieve thermal equilibrium, then the temporal approach to equilibrium might be observed.

In the case of materials with strong chemical bonds of a few eV, thermal equilibration depends on shock pressure, density, shock energy deposited, bond strengths and experimental life time. The lowest shock pressure in a strong oxide ($Gd_3Ga_5O_{12}$ (GGG)) at which thermal equilibrium has been observed to occur in a narrow shock front on the Hugoniot is 130 GPa (Zhou et al., 2015). Shocked GGG is in thermal equilibrium for $P_H > 130$ GPa. GGG is the only strong oxide to date for which the minimum P_H that rapidly induces thermal equilibrium has been determined experimentally, to the knowledge of this author.

While thermal equilibrium is not possible in all materials during experimental lifetime, neither is it essential in all dynamic experiments. In many strong materials, shock-induced *mechanical equilibrium* is often sufficient. Sapphire (single-crystal Al_2O_3) is used as anvil in multiple-shock experiments. In this case, shocked sapphire need only be in mechanical equilibrium so that reproducible multiple-shock states in fluid hydrogen are achieved, for example. In this case, the sample (liquid H_2) rapidly equilibrates thermally in the fluid phase, while sapphire equilibrates mechanically below $P_H \approx 130$ GPa and probably equilibrates thermally as well for $P_H > 130$ GPa, at which shocked sapphire becomes opaque (Urtiew, 1974).

An extensive body of experimental EOS data under extreme dynamic shock pressures has been measured (Marsh, 1980; Trunin, 1998, 2001). Those results are used to design dynamic experiments. Trunin's (2001) compendium of experimental data up to 20 TPa (200 million bar) is unique and extremely valuable for learning systematic "universal" behavior of materials at extreme shock pressures from few 0.1 TPa up to 20 TPa (Chapter 10).

1.5 Recent Accomplishments

Dynamic compression achieves ultracondensed matter at extreme conditions, which offers opportunities to understand phenomena not understood previously and to discover new interesting phenomena in regimes studied little, if at all. Measured dynamic-compression data are relevant to physics, chemistry, materials science, geophysics, planetary science, exoplanets, astrophysics, Inertial Confinement Fusion (ICF), etc. Dynamic compression thrives on the creativity of its users, and its uses are limited only by the imaginations of its practitioners. The following sections provide examples in which dynamic compression has been used to address questions of long-standing scientific and technological interest and to point out opportunities for future research.

1.5.1 Metallic Fluid Hydrogen (MFH)

High-pressure researchers have long been motivated to make metallic hydrogen whose existence was predicted by Wigner and Huntington (1935). With the end of the Cold War in 1989, dynamic high-pressure researchers in defense laboratories started investigating the possibility of making MFH with dynamic high pressures, because at that time metallic hydrogen had yet to be made under static high pressures. Degenerate metallic hydrogen cannot not be made with single-shock compression, because shock-heating limits shock compression to a value too small to make metallic H by quantum-mechanical overlap of electronic wave functions on adjacent H atoms. Dissipation T and S tuned by time history of a pressure pulse applied to liquid H_2 made degenerate (metallic) fluid H by dissociation of H_2 to H at sufficiently high pressures and densities and low temperatures. While temperatures exceeded melting temperatures at those pressures, those temperatures were sufficiently low to produce degenerate metallic fluid H.

WH's predicted metallization density, 0.62 mol H/cm^3, corresponds to a static pressure of 73 GPa at 300 K (Loubeyre et al., 1996). Temperature and entropy generated by dynamic compression achieved at finite temperatures with a two-stage light-gas gun (2SG) drove a crossover via dissociation from H_2 to H which completes at 0.63 mol H/cm^3 (3.8×10^{23} H/cm^3) and 140 GPa to make degenerate MFH at 140 GPa, T = 3000 K, T_F = 220,000 K and $T/T_F \approx 0.014 \ll 1$. That crossover completes essentially at the density of metallization of the first-order insulator-metal transition predicted by WH (1935). MFH under dynamic compression was achieved by reverberating a shock wave in liquid H_2 contained between two anvils of strong dense sapphire (single-crystal Al_2O_3) (Nellis, Weir and Mitchell, 1999), as discussed in Chapter 6. The key enabler of this discovery was recognizing that dynamic compression is not restricted to compression by a single, sharp shock front.

In contrast, at a temperature of 5.5 K in diamond anvil cell (DAC) a sample of metallic solid (probably) H (MSH) has recently been reported at a static pressure of 495 GPa (Dias and Silvera, 2017). Although an absolute static pressure scale above ~200 GPa has yet to be developed, the pressure of 495 GPa was determined by linear extrapolation of a diamond-Raman pressure scale (Akahama and Kawamura, 2007). The plasma frequency of MSH derived from reflectance data corresponds to an electron density of 6.6 to $8.8 \times 10^{23}/cm^3$, which agrees with estimates of H atom density and is twice the electron density estimated above for MFH. Metallization pressure and density of MFH and MSH differ strongly between 140 GPa/3000 K and 495 GPa/5.5 K for fluid and solid, respectively.

1.5.2 Unusual Magnetic Fields of Uranus and Neptune: Metallic Fluid H

In the 1980s, NASA's Voyager 2 spacecraft measured the unusual non-dipolar non-axisymmetric magnetic fields of the Ice Giant Planets Uranus and Neptune (U/N), whose magnetic fields are unique in the solar system. The reason why those fields differ so much from other known planetary magnetic fields has been a scientific puzzle ever since their discovery. A picture of the interiors of U/N that might cause those unusual field geometries should be consistent with (1) measured properties of fluids likely to be in U/N at likely pressures and temperatures at which those magnetic fields are made by convective dynamo motion and with (2) density distributions of U/N determined by gravitational data measured by the Voyager 2 spacecraft. Because interior compositions and thermodynamic conditions in U/N are not known in detail, a reasonable approach is to measure on Earth a substantial property database to look for systematics which, together with gravitational data of U/N measured by Voyager 2, might provide a reasonable explanation of those fields, independent of specific interior compositions and interior P and T at which those magnetic fields are made. This has in fact been done.

Numerous electrical conductivity and EOS measurements have been made on likely planetary molecular liquids to accumulate an extensive experimental database at high pressures and temperatures that might be useful in developing an explanation of the Ice Giants (Nellis, 2015b). When those measurements on Earth began around ~1980 (Nellis et al., 1981, 1983; Mitchell and Nellis, 1982), no one had any idea what Voyager 2 would find when it arrived at U/N in the late 1980s. That study at Lawrence Livermore National Laboratory (LLNL), a national defense laboratory, persisted more than ~30 years out of intellectual curiosity, justified by the fact that molecular liquids thought to have accreted to form U/N are also likely constituents of chemical explosives. Thus, experimental investigation of such liquids under dynamic compression is effectively a study of reaction products of detonated chemical explosives, which gave that experimental program an applied technology rationale. That long-term scientific study has provided the great majority of experimental data collected systematically on likely planetary fluids at high *P/T* thought to be in U/N.

Based on that extensive database measured on Earth, particularly the measured pressure dependence of the unexpectedly large electrical conductivities up to 2000/(Ω-cm) of dense fluid hydrogen as it crosses over from semiconducting to metallic fluid H (Weir et al., 1996a; Nellis et al., 1999) and the radial-density distributions of U/N determined by Voyager 2 (Helled et al., 2011), the unusual magnetic fields of U/N are probably made relatively close to their outer radii primarily by convection of MFH. Likely nebular planetary molecular species, called "Ices", accreted to form U/N decompose at *P* and *T* estimated in the Ice cores of U/N. Because of this decomposition, Ice cores of U/N are expected to be chemically complex mixtures of decomposed nebular gases, silicates and Fe/Ni alloys. Convection in those cores probably contributes to the magnetic fields of U/N (Nellis, 2015b), but magnetic-field contributions from the Ices are not the dominant contributions to the magnetic fields of U/N, as has been suggested incorrectly (Nellis et al., 1988a). The likely cause of the complex spatial shapes of those magnetic fields relative to Earth's magnetic field is decoupling of planetary rotational motion from convective dynamo motions, which generates those magnetic fields. In addition, it was scientifically important to learn how the deep-Earth geophysics that causes the nearly dipolar and axisymmetric magnetic field of Earth does not exist in U/N, and so it is quite likely that rotational motions of U/N are decoupled from convection in MFH, which makes the unusual magnetic field shapes of U/N (Chapter 7).

1.5.3 General Issues in Planetary Science

In recent years a number of experiments and computational models have been done that are important for general issues in planetary science. Dynamic compression achieves representative high *P/T* states at which physical and chemical properties of planetary materials are measured and enables hypervelocity phenomenological experiments relevant to evolution of planets and moons in the solar system. These include (1) EOS measurements on geological oxides (Luo et al., 2004) and water ice (Stewart et al., 2008), (2) phenomenology of crater formation on surfaces of icy moons, including scaling laws for melting, vaporization and crater size (Kraus et al., 2012, 2015) and the possible importance of ascending thermal plumes in the formation of such craters (Medvedev, 2010), and (3) survival, destruction and formation on hypervelocity impact of organic molecules and yeast spores (Bowden et al., 2009; Furakawa et al., 2009; Price et al., 2013). These are relevant to the question of how life is transported between bodies in the solar system and beyond.

1.5.4 General "Atomic" States of Warm Dense Fluids at Extreme Conditions in Exoplanets

Based on experiments and theory to date, fluids at extreme conditions achieved by (1) QI multiple-shock compression of H, N and O at dynamic pressures up to 0.18 TPa and a few 1000 K (Chau et al., 2003a) and Li (Bastea and Bastea, 2002), (2) fluid alkali metals Rb and Cs under static pressures of 10^{-5} TPa on saturation curves near 2000 K (Hensel and Edwards, 1996; Hensel et al., 1998; Edwards, et al., 1998), (3) fluid SiO_2 under single-shock pressures up to 1.6 TPa (Knudson and Desjarlais, 2009) and fluid $Gd_3Ga_5O_{12}$ (GGG) under shock pressures up to 2.6 TPa (Ozaki et al., 2016), (4) fluid Al, Cu, Fe and Mo at shock pressures up to 20 TPa (Trunin, 2001; Nellis, 2006b; Ozaki et al., 2016), (5) fluid Xe (Root et al., 2010) and fluid Kr (Mattesson et al., 2014) up to 0.9 TPa and (6) theoretical calculations of the Hugoniot and conductivity of Al_2O_3 up to ~1.5 TPa and 15,000 K (Liu et al., 2015) are observed to have or are expected to reach minimum metallic conductivity (MMC) of few 10^5 S/m in a common end state that is probably an "atomic-like" fluid. That common end state is warm dense matter (WDM) with electron correlations. Systematics in that data are potential experimental bench marks for developing theories of WDM.

Above 0.3 TPa on the Universal Hugoniot of fluid metals (UHFM), measured Hugoniot and optical reflectivity data indicate strong $Gd_3Ga_5O_{12}$ (GGG) at ambient crosses over to a semiconductor less compressible than diamond above 170 GPa and then to a compressible, poor metal (Ozaki et al., 2016). Theoretical electron-band calculations of disordered GGG indicate such a crossover in the same pressure range (Ozaki et al., 2016). That is, weakening of interactions in strong, metal-oxygen bonds between atoms at sufficiently extreme dynamic pressures and temperatures drives a crossover to itinerant electrons in a compressible fluid metal, an idea analogous to the concept of asymptotic freedom in particle physics. That is, quarks and gluons bind strongly at ambient as indicated by nuclear stability. However, quarks in nuclei interact more weakly at high energies as observed by the fact that perturbation theory calculates deep inelastic scattering cross sections at sufficiently high energies (Gross and Wilczek, 1973; Politzer, 1973).

1.5.5 Strong Transparent insulators and Shock-Induced Opacity

Metallic fluid H, N and O were made in the range 100–140 GPa by reverberating a shock wave in the diatomic liquid contained between two strong electrically insulating sapphire anvils. In this case, relevant physical properties of sapphire in the range 90 to 180 GPa must be known to interpret measured experimental data of those fluids. Prior to those experiments relatively few properties of strong oxides had ever been measured above a shock pressure of 90 GPa.

Sapphire was chosen for those anvils because the Hugoniot of sapphire had already been measured by McQueen et al. up to 140 GPa (Marsh, 1980, pp. 260–261) and by Erskine (1994) up to 340 GPa, which meant dynamic compression experiments on diatomic liquids using sapphire could be designed. However, electrical resistivities of sapphire also needed to be measured up to ~200 GPa shock pressures to determine the range of shock pressures in which electrical current would not be shunted through shocked sapphire rather than transported through shock-compressed fluid hydrogen. For this reason, electrical conductivities of shocked sapphire were measured in the range 90 to 220 GPa (Weir et al., 1996b) and found to be negligible with respect to conductivities of metallic fluid H, N and O at 140 GPa. Sapphire anvils in those shock reverberation experiments were electrical insulators.

Temperatures of MFH above ~100 GPa shock pressures cannot be measured because shock-compressed sapphire becomes opaque via shock-induced defects in the range 100 to 130 GPa (Urtiew, 1974). Thus, gray-body temperatures of MFH cannot be determined from thermal spectra emitted from warm dense hydrogen, which spectra must pass through shocked sapphire. Thus, it is not yet possible to measure Planck thermal spectra emitted from MFH and thus it is not yet possible to measure temperatures of MFH that are needed to develop theory of MFH.

To interpret conductivity data measured in shock reverberation experiments, it was necessary to verify that strength effects in sapphire are negligible above 90 GPa so that such effects could be neglected in data analysis. For this reason shock-wave profiles were measured for shock waves propagating in seven crystallographic directions in the sapphire rhombohedral lattice (Kanel et al., 2009). Those wave-profile results verified that strength effects in sapphire are negligible at shock pressures above 90 GPa. It is possible, however, that at least one of the Al_2O_3 crystal orientations other than *c*-cut might provide improved optical properties relative to those of *c*-cut sapphire (Kanel et al., 2009; Liu et al., 2015).

To vary P, ρ and T achieved in fluid hydrogen with shock reverberation, crystals with various densities and shock impedances other than those of sapphire need to be characterized. For this reason GGG has been investigated extensively in Hugoniot measurements (Mashimo et al., 2006), shock-wave profiles (Zhou et al., 2011) and optical-emission measurements (Zhou et al., 2015) in a search for a strong oxide that might be transparent at shock pressures above 140 GPa.

Unfortunately, GGG becomes opaque at a shock pressure of 130 GPa, as does sapphire. Shock-induced opacity in GGG is caused by optical absorption and scattering as determined with measured 16-channel optical spectra (Zhou et al., 2015). Nevertheless those GGG experiments provide the spectrographic signature of shock-induced optical opacity, which provides an important clue in the search for crystals that remain transparent above shock pressures of 140 GPa, so that thermal emission spectroscopies of MFH can be measured.

1.5.6 Inertial Confinement Fusion (ICF): Deuterium-Tritium Fuel

ICF, a technological application of dynamic compression, is currently under development with quasi-isentropic multiple-shock compression (Nuckolls et al., 1972; Motz, 1979; Lindl et al., 2014; Hurricane et al., 2014), analogous to the QI technique used to make MFH. The purpose of this effort is to develop an alternative source of commercial energy. Current ICF fuel is molecular deuterium-tritium (D_2-T_2). Metastable solid metallic H in the form of solid D-T would provide a substantially higher fuel density for ICF, if solid metallic D-T could be made metastably at ambient. Energy produced by thermonuclear burn of metastable metallic D-T fuel targets is expected to be substantially higher than that of significantly less-dense solid molecular D_2-T_2 targets. In this case it would be the high density of solid metallic D-T that would be attractive, rather than its metallic nature.

In addition, solid metallic D-T fuel would have a higher mass density than molecular D_2-T_2 and thus might also have a relatively small difference in density relative to its encapsulation material in the fuel ball. Because it is the density difference between D-T fuel and its encapsulation material that drives growth rates of the Richtmyer-Meshkov (R-M) (Mikaelian, 1985) and Rayleigh-Taylor (R-T) (Strutt, 1883; Taylor, 1950) hydrodynamic interfacial instabilities during capsule implosion, a higher mass density of D-T fuel might reduce growth rates of those hydrodynamic interfacial instabilities. In this case, significantly higher ICF fusion energy yields would also be expected. Use of metastable metallic D-T fuel pellets offers the prospect of substantially higher ICF fusion-energy yields, and thus the prospect of a substantially higher probability of the success of the ICF Program. Given that the ICF Program at LLNL began ~1970 and is yet to produce practical fusion energy, the time is more than right to attempt to develop metastable solid metallic D-T fuel pellets to achieve higher ICF energy yields.

1.5.7 Potential Novel Materials: Metastable Solid Metallic Hydrogen (MSMH) and Others

Dynamic compression offers a possible way to quench fluids with attractive properties at extreme conditions to solids at ambient for scientific and technological applications. While many such possibilities exist, metastable solid metallic hydrogen (MSMH) probably has the most to offer in this regard, in addition to a potential ICF fuel. Possible scientific applications, if MSMH could be made, are a quantum solid at room temperatures with potentially novel physical properties, including room-temperature superconductivity. Possible technological applications might be a light-weight (~1 g/cm^3) structural material and a more energetic fuel or propellant depending on rate of release of stored energy (Nellis, 1999). Figuring out how to make MSMH with all its important potential scientific and technological applications is a problem in which solutions have substantial implications for Research and Development and for economies on a national and international scale.

1.6 Bibliography

Design and interpretation of experiments at extreme dynamic conditions involve supersonic hydrodynamics, thermodynamics, condensed matter physics and chemistry, materials science, planetary science, geophysics, astrophysics, ICF, and so forth. Dynamic compression is a technique that offers a wide array of potential discoveries in a wide variety of multiple disciplines that is limited only by the imaginations of its practitioners. A bibliography follows on a variety of topics involving dynamic compression. Previous reviews on dynamic compression deal with single-shock compression: Courant and Friedrichs, 1948; Rice et al., 1958; Altshuler, 1965; Zeldovich and Raizer, 1966; Kormer, 1968; McQueen et al., 1970; Duvall and Graham, 1977; Davison and Graham, 1979; Marsh, 1980; Melosh, 1989; Bushman et al., 1992; Asay and Shahinpoor, 1993; Cheret, 1993; Horie and Sawaoka, 1993; Graham, 1993; 1994; Meyers, 1994; Trunin, 1998; 2001; Wilkins, 1999; Nellis, 2002a; Fortov et al., 2004; Kanel et al., 2004; Ahrens, 2005; Isbell, 2005; Zhernokhletov, 2005; Nellis, 2005; 2006a; Hicks et al., 2009; Eakins and Thadhani, 2009; Nellis, 2010; Dlott, 2011; Lemke et al., 2011; Forbes, 2012; Kraus et al., 2012; Knudson and Desjarlais, 2013; Lindl et al., 2014; Krehl, 2015; Nellis, 2015a; 2015b; Banishev et al., 2016; Fortov, 2016; Asay et al., ed., 2017.

2

Basics of Dynamic Compression

Dynamic compression developed as part of classical physics in the last half of the nineteenth century. The dynamic compression process is based on conservation of mass, momentum and energy and associated generation of dissipation in the form of temperature and entropy. Because thermodynamics can be tuned to a great degree, dynamic experiments can be designed to probe specific scientific issues in specific regimes of P and T. While dynamic compression itself is generally a part of classical physics, states of matter produced by the process range from quantum-mechanically degenerate to non-degenerate, depending on density ρ and temperature T achieved by the process.

Supersonic nonlinear dynamic compression involves shock compression, shock-pressure release, transmission and reflection of shock waves at interfaces between two materials and reverberating and intersecting dynamic waves. For this reason dynamic compression needs to be handled with nonlinear hydrodynamic computational simulations. Supersonic hydrodynamics is a branch of pure and applied mathematics (Liu, 1986). A brief description of equations that govern supersonic flows is given in Section 2.1.16.

A paradigm effect of nonlinearity is the case of two 1-D compressional shock waves, equal in magnitude, that approach and pass through each other in opposite directions. The result of two such overlapping shock waves is a tensile wave (*negative pressure*) that might spall a material into two separate layers (Wilkins, 1999). That result is not an increase of compressible shock pressure, as might be expected for linear waves. It is important to know how to avoid or to use nonlinear effects, and not be surprised by them.

The purpose of this chapter is to discuss basic aspects of dynamic compression to facilitate design of experiments and interpretation of affects. Knowledge presented herein is useful, for example, for generating a design of an experiment with "back-of-an envelope" calculations in the spirit of Fermi's approach to physics (von Baeyer, 1993; Cronin, 2004). Once a basic design is determined, nonlinear computational simulations can be performed to verify that desirable effects are achieved and undesirable effects are not.

The final state achieved by dynamic compression depends on the temporal history of the pressure pulse applied by a matter wave. For example, supersonic shock compression of a liquid occurs by a single fast jump in pressure to a thermally equilibrated final state behind the front of that matter wave. A shock wave is one particular type of dynamic-compression wave. The locus of states achieved by a sequence of single-shock compressions, each from the same initial density, is commonly called a Hugoniot curve or simply a Hugoniot. Dynamic isentropic compression occurs virtually at the speed of sound, which is the limit of an infinite number of infinitely weak shocks (Zeldovich and Razier, 1966). Shock compression is substantially faster and generates substantially higher temperatures and entropy than does quasi-isentropic compression. Quasi-isentropic states are intermediate between those obtained by shock and isentropic compression and are tuned by an effective combination of shock and isentropic compression (Nellis, 2006a).

Dynamic compression is adiabatic because supersonic velocities of the compressive wave front are greater than longitudinal sonic velocities of thermal transport in the direction parallel to wave propagation. Similarly, in a finite-size sample, velocities of the compressive wave front are greater than sonic velocities of thermal transport out of the thin wave front in the direction perpendicular to that of wave propagation. As a result, a central region of a sample survives unaffected by edge effects for some length of time. It is such a region that is generally diagnosed. Layer thicknesses in a sample must be chosen such that the unaffected region is sufficiently large during experimental lifetime that an accurate measurement can be made with available experimental time resolution.

Dynamic compression is dissipative because equilibrated energy, in the form of temperature T and equilibrated shock-induced disorder entropy S, are not available for compression. Dissipation energy $E_d = TS$ is shock energy absorbed by mechanisms other than compression. As discussed in Chapter 1, T and S produced dynamically are tunable over substantial limits, which enables tuning off Hugoniot experimental conditions over a substantially larger thermodynamic range than possible to access with single-shock compression alone. The realization that dynamic compression is not restricted to a Hugoniot curve is arguably one of the most significant advances in dynamic compression in decades.

In practice, the terms "dynamic compression" and "shock compression" are often used interchangeably because supersonic compression is a key feature of both. Other common examples of dynamic compression include dynamic isentropes and multiple-shock waves. A multiple-shock wave is a sequence of several individual shock jumps, which together produce quasi-isentropic compression. The first jump is a shock wave, which produces significant shock dissipation; the ensemble of sufficiently many successive shock jumps forms an isentrope from that first-shock state.

A ramp wave is often an initial relatively weak jump in shock pressure followed by a ramp from the initial shock state up to final pressure. Final pressure and temperature are tuned by choice of initial shock amplitude and rate of increase of pressure with time. Temperatures and densities produced by a ramp wave are quasi-isentropic and thus lower and higher, respectively, than those produced by single-shock compression.

2.1 Shock Compression

In this work we are interested primarily in general conceptual ideas of experimental configurations, rather than technical details specific to particular shock drivers. To present a context for these discussions, experiments are discussed primarily with respect to those performed with dynamic waves generated with a two-stage light-gas gun. Experiments performed with other generators of dynamic compression waves are similar in concept though different in temporal and spatial details. Once conceptual ideas are understood, they can readily be transferred to corresponding concepts appropriate for other shock generators.

We begin with a discussion of a generic geometric configuration typical of most shock-compression experiments, illustrated in Fig. 2.1. This configuration is of the general stimulus/response type of experiment. The impactor generates a shock wave in the forward direction in the left wall of the liquid Ar sample holder. The impact shock transits that wall and eventually crosses the boundary with Ar. Shock breakout into the liquid generates a shock wave in the forward direction in liquid Ar with a different pressure than the initial shock pressure in the wall. On breakout,

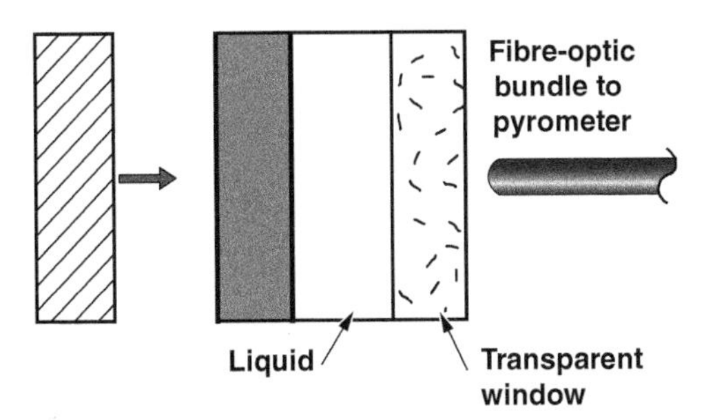

Fig. 2.1. Schematic illustration of shock compression experiment to measure optical emission. Impactor on left generates P, ρ, T state in liquid. When impact shock breaks out of opaque wall (gray) on left of sample holder, shock front in liquid emits thermal radiation spectrum. (Nellis et al., 1997). Copyright 1997 by American Institute of Physics.

the shock rapidly compresses and heats the Ar, which generates a fast optical response from the shock front in the heated dense fluid. Emitted thermal radiation passes through transparent liquid Ar ahead of the shock front and then through the window on the right. Emitted optical radiation is then diagnosed with a time-resolved multi-channel pyrometer or spectrograph (not shown) on the right side of the sample holder opposite the impact. Effective emission temperature of the shock front can be determined if the emission spectrum is gray-body, that is, a Planck spectrum with emissivity ε, such that $0 < \varepsilon < 1$.

The impactor could be driven by chemical explosives, by a two-stage light-gas gun, by pulsed magnetic pressure on a metal impactor plate or by a pulsed laser. Various detectors could be used to measure properties other than thermal emission spectra. Other possible properties include shock velocity to determine Hugoniot equation-of state points (P_H, ρ_H, E_H) and electrical conductivities. Whatever the shock generator and detector, configurations of experiments are usually conceptually similar to Fig. 2.1, while technological details depend on the shock driver.

To design and interpret such experiments, it is essential to have a basic understanding of shock phenomena. Common questions include (1) what shock pressure is achieved in a given impact, (2) how does shock pressure change on transit by a shock wave across a boundary between two materials, (3) is a given shock-compressed sample in thermal equilibrium, (4) which sample dimensions are required such that shock compression is one-dimensional during experimental lifetime and (5) what sample dimensions are required such that diagnostic signals can be measured accurately with available diagnostic time resolution?

2.1.1 Simple Shock Front on an Atomic Scale

Atoms in liquid Ar (Fig. 2.1) interact via effective van der Waals pair potentials, which are relatively weak and describe a compressible shock-heated liquid. A simple shock front in liquid Ar is illustrated in Fig. 2.2 on (a) macroscopic and (b) microscopic spatial scales. The shock front is the thin (~nm) transition region in (b) in which Ar atoms are rapidly compressed and equilibrate thermally via inter-atomic collisions with associated turbulent frictional heating to T and generation of associated entropy S within the thin shock front. Values of thermodynamic parameters behind a shock front with H subscripts are uniquely determined by their values ahead of the shock front with subscripts 0, independent of the path taken toward thermal equilibrium in the extremely non-equilibrium conditions in the width of a shock front (Bethe and Teller, 1940). *Turbulent heating and entropy generated within a thin shock front are the essence of dynamic compression.*

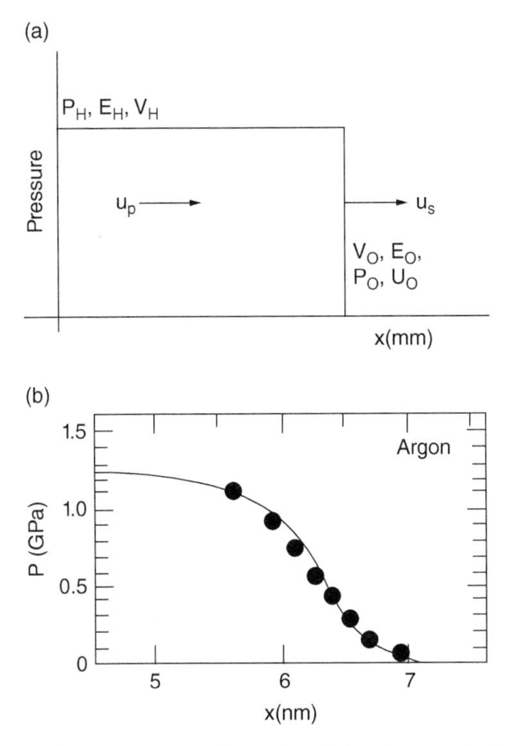

Fig. 2.2. (a) Schematic on macroscopic scale of shock wave in liquid. P_H, E_H and V_H ($1/\rho_H$) are state variables of shock pressure, internal energy on shock compression and shock volume, respectively; zero-subscripted variables are initial values ahead of shock front; u_s and u_p are hydrodynamic flow variables of supersonic shock velocity and subsonic material velocity behind shock front, respectively. (b) (Same as **Fig. 1.1**) Calculated shock-front width showing rise in pressure in Ar to 1.2 GPa in ~nm. Shock speed is 1.8 km/s; rise time from 0 to 1.2 GPa is ~10^{-12} s. Apparent smoothness of rise is caused by spatial averaging in non-equilibrium shock front. Full curve calculated with atomistic molecular dynamics (Klimenko and Dremin, 1979); solid circles calculated with Navier-Stokes equations (Hoover, 1979). Calculated temperature is ~500 K at 1.2 GPa. Copyright 1979 by American Physical Society.

Relative amounts of shock dissipation T and S depend on the nature of interactions between atoms. Calculated rise time of a 1.2 GPa (12 kbar) shock wave in liquid Ar in Fig. 2.2b is ~10^{-12} s (Klimenko and Dremin, 1979; Hoover, 1979). The shock front has a spatial width of ~nm, which in Ar corresponds to ~4 atomic Ar diameters. The rise time of ~10^{-12} s is many orders of magnitude faster than the rise time of static pressure in a DAC of ~s.

Because of the thin (nm) shock front in Fig. 2.2b, pressure jumps virtually discontinuously from P_O at V_O to P_H at V_H. The Rayleigh line is a straight line in P-V space drawn directly between (V_O, P_O) and (V_H, P_H), which emphasizes the

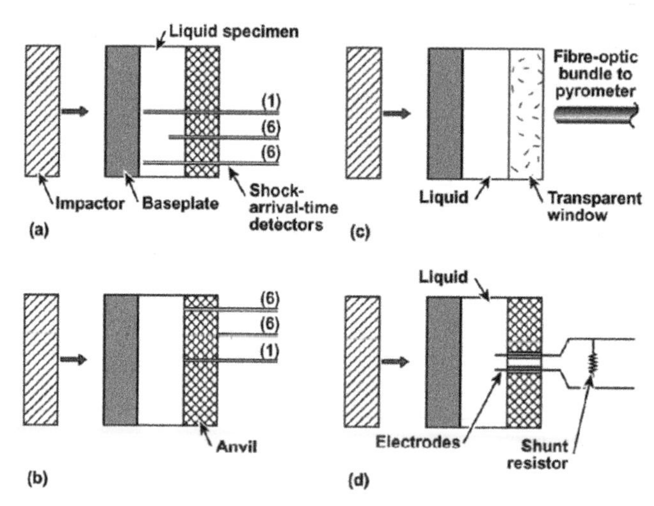

Fig. 2.3. Illustrations of single-shock experiments in liquid to measure (a) velocity of shock wave with discrete shock-arrival time detectors to determine Hugoniot equation-of-state point, (b) shock velocity in anvil to determine double-shock state, (c) spectral dependence of optical radiation emitted from shock front to determine gray-body temperature (same as Fig. 2.1), and (d) electrical conductivity. Numbers in parenthesis in (a) and (b) are numbers of point detectors on same plane. Velocity of impactor is measured in-flight with flash x-radiography (Nellis et al., 1997). Copyright 1997 by American Institute of Physics.

sharp jump from initial to final states. In contrast, on isotherms and isentropes, P increases continuously with decreasing volume V.

Several types of single- and double-shock experiments on liquid specimens performed with impact-generated shock waves are illustrated schematically in Fig. 2.3. "Planar" shock fronts generated by impacts with plates accelerated with a 2SG are generally tilted and distorted. For Hugoniot experiments, which measure shock velocity, the spatial shape of the front of a shock wave is diagnosed to determine how to optimally average measured shock arrival times. Shock waves generated with a 2SG are generally tilted $\leq 1^0$. In addition the shock front can be curved parabollically to a typical depth of ~20 μm over a sample radius of ~12 mm (Nellis and Mitchell, 1980; Mitchell and Nellis, 1981a).

Single- and double-shock experiments to measure shock velocities in liquids are illustrated schematically in Figs. 2.3a and 2.3b (Nellis and Mitchell, 1980; Nellis et al., 1983). Detectors in the circular arrays in Figs. 2.3a and 2.3b measure average arrival times on two levels separated by 1–2 mm. The center pin on axis enables measurement of the difference between average apparent shock arrival time on axis of a 6-detector circular array positioned on a radius of 12 mm and measured arrival time on axis. This interval time defines the parabolic distortion of the impactor plate during its acceleration to $\leq$ 8 km/s along the ~10 m length of the launch tube of the two-stage gas gun (2SG). Figure 2.3c is identical to Fig. 2.1 above. Fig. 2.3d illustrates an experiment to measure electrical conductivity of shock-compressed water (Mitchell and Nellis, 1982).

To achieve highest pressures and lowest temperatures under shock compression, specimens should have the highest initial density (see Eq. (2.1)). For example, in the case of Ar and H_2, the normal boiling points are 87 K and 20 K, respectively. Thus, high initial density can be achieved by using either a condensed cryogenic liquid specimen at atmospheric pressure or a dense gas at high pressure and room temperature. Experiments performed with a 2SG at LLNL have used cryogenic liquid specimens (Nellis and Mitchell, 1980; Nellis et al., 1983) because (1) the complexity introduced by cryogenics was less than complexity required with high-pressure gas specimen holders and (2) accurate equations of state near atmospheric pressure are available for cryogenic liquids on their saturation curves. Thus, accurate densities of liquid specimens are readily obtained simply by measurements of barometric pressure.

Illustrations in Fig. 2.3 describe the basic concepts of such diagnostics. Over the decades these basic concepts have remained virtually unchanged. However, detector technology has developed enormously over recent decades with respect to detection mechanism, sample size, experimental lifetime and temporal and spatial resolutions. Details involved in using other detection systems, for example, measuring shock arrival times continuously across a sample diameter rather than pointwise around a circular array, are available in the scientific literature.

2.1.2 Rankine-Hugoniot Equations

The R-H equations conserve momentum, mass and internal energy across the front of a shock wave (Fig. 2.2b). They relate initial state variables ahead of a travelling shock front, denoted by zero subscripts in Eqs. (2.1) to (2.3), to shock-state variables behind the shock front, denoted by H subscripts, and to flow variables u_s and u_p, where u_s and u_p are supersonic shock velocity of the shock front and subsonic velocity of material behind the shock front, respectively. The R-H conservation equations are

$$P_H - P_0 = \rho_0 (u_s - u_0)(u_p - u_0), \tag{2.1}$$

$$V_H = V_0\left[1 - (u_p - u_0)/(u_s - u_0)\right], \tag{2.2}$$

$$E_H - E_0 = 0.5(P_H + P_0)(V_0 - V_H), \tag{2.3}$$

where mass density $\rho_H = 1/V_H$, $\rho_O = 1/V_O$, and u_O is initial particle velocity in front of the shock wave. V_H and E_H are specific volume and specific internal energy, respectively, where "specific" means per gram g. Initial sample temperature T_0 enters implicitly through ρ_0 and E_0. For solids with strength, pressure P is replaced by stress σ_n in the direction of shock propagation. The R-H equations describe an adiabatic process in which zero thermal energy is transported into or out of the shock front during compression. Internal energy E_H is deposited only by shock compression itself. Entropy S increases under dynamic compression such that $S_H \geq S_0$, where S_H and S_0 are shock and initial entropies, respectively (Courant and Friedrichs, 1948). The R-H conservation equations apply only across the front of a single shock wave (Krehl, 2015).

Eqs. (2.1) to (2.3) show that by measurements of u_s and u_p, or by measurement of u_s and determination of u_p by a process called shock-impedance matching, P_H, V_H and E_H are obtained. Velocities u_s and u_p are obtained from measurements of shock transit time over a pre-measured distance and from initial thermodynamic state (ρ_0) of the sample.

The locus of states achieved by a sequence of shock compressions, each from the same density, is called a R-H curve, a Hugoniot, or a shock adiabat. The R-H conservation Eqs. (2.1) to (2.3) apply whether or not shock-compressed material is equilibrated thermally. The question of thermal equilibrium must be decided independent of the R-H conservation equations. In this work, entropy in shock-compressed fluids is thermally equilibrated disorder S by virtue of high shock temperatures, generally greater than ~1000 K herein.

In most materials u_s and u_p are related linearly in significant ranges of u_p as

$$u_s = C + Su_p, \tag{2.4}$$

where C and S are constants, as illustrated in Fig. 2.4a. Ideally, C is the speed of sound at ambient and B_0^S is isentropic bulk modulus at ambient and $S = \left(B_0^{S'} + 1\right)/4$, where $B_0^{S'}$ is the pressure derivative of isentropic bulk modulus at ambient (Ruoff, 1967). Eqs. (2.1) and (2.4) show that at high shock pressure $P_H \propto u_s u_p$, $\propto u_p^2$, as illustrated in Fig. 2.4b. P_H and u_p are the only shock variables in the R-H equations that are continuous across a boundary between

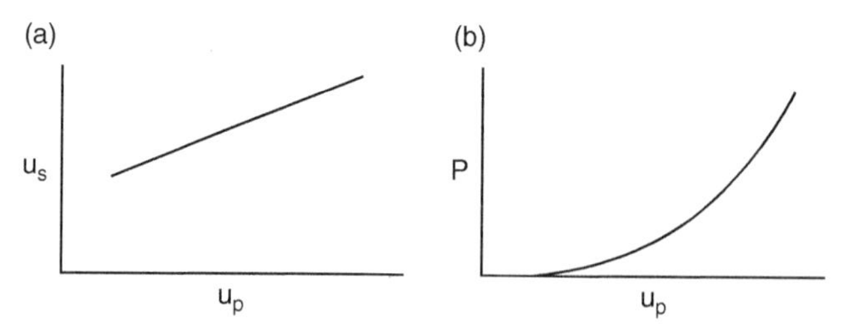

Fig. 2.4. (a) Shock and particle velocities u_s and u_p, respectively, are related linearly in significant ranges of u_p. (b) For large u_p, $P_H \propto u_p^2$.

two different materials. For this reason $P_H(u_p)$ curves are used extensively in calculating shock-compression states achieved on impact and on shock propagation across boundaries between different materials using a technique known as shock impedance matching.

The R-H Eqs. (2.1) and (2.2) can be solved for u_s and u_p as functions of P_H and V_H. Shock velocity u_s is given by

$$u_s = V_0[(P_H - P_0)/(V_0 - V_H)]^{1/2} = V_0\left[S_{ray}\right]^{1/2}, \tag{2.5}$$

where S_{ray} is the absolute value of the slope of the Rayleigh line. Particle velocity u_p is

$$u_p = [(P_H - P_0)(V_0 - V_H)]^{1/2}. \tag{2.6}$$

A phase change is often accompanied by a volume change that causes a change in slope of $P_H(V_H)$, which is also observed as a change in slope of $u_s(u_p)$. Conversely, if a phase change is not accompanied by a volume change, neither does a slope change occur in $u_s(u_p)$.

These equations for $u_s(u_p)$ are particularly relevant for Hugoniot measurements of metals made at ultrahigh pressures above ~ TPa. Theoretical predictions of such Hugoniots with unusual shapes and curvatures have been made in P_H (V_H) space (Rozsnyai et al., 2001). Such Hugoniots have also been measured underground in $u_s(u_p)$ space (Rozsnyai et al., 2001). Systematic analysis of measured Hugoniot data up to 20 TPa show Hugoniots of common metals essentially have a common linear dependence in $u_s(u_p)$ space (Nellis, 2006b; Ozaki et al., 2016), the space in which Hugoniot points at ultrahigh pressures are usually measured. It would be useful for experimentalists if theoretical predictions of unusual Hugoniot shapes at ultrahigh pressures were also published in $u_s(u_p)$ space as a guide to sensitivity and accuracy required in measurements of shock velocity.

2.1.3 Shock-Pressure Release

Shock dissipation energy, $E_d = TS$, is irreversible. Because liquid Ar is disordered initially, most E_d in shock-compressed fluid Ar goes into increased T above initial temperature T_0 and relatively little S goes into increased S above initial value S_0. On release of shock pressure to zero, because of irreversible E_d, the sample does not return to its initial state. Rather, on release of shock pressure P_H to 0, the sample has a residual temperature $T_{res} > T_0$ and residual density $\rho_{res} < \rho_0$ by virtue of irreversible shock-heating retained on expansion to ρ_{res}. Shock pressure releases essentially at the speed of sound, which is much slower than the rise time of a supersonic shock front. Thus, shock pressure release from very high pressures is

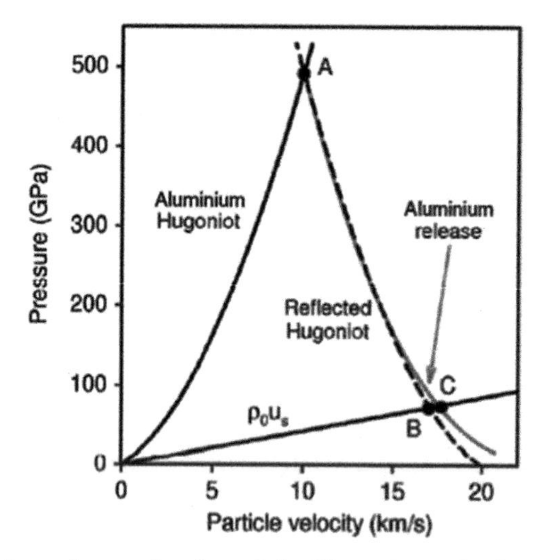

Fig. 2.5. Hugoniot, mirror reflection of the Hugoniot and release isentrope from Hugoniot of Al at point A at 500 GPa (Knudson et al., 2004). Points B and C are on mirror reflection and release isentrope, respectively. Strait line is Hugoniot relation of liquid H_2 for pressure: $P_H = \rho_0 u_s u_p$. Copyright 2004 by American Physical Society.

virtually isentropic, except at the very lowest release pressures at which residual effects of $E_d = TS$ deposited at highest shock pressure are significant. Pressure release curves are release adiabats and commonly called release isentropes.

Consider an Al layer shock compressed to point A on its Hugoniot in Fig. 2.5. Consider further that the Al layer is backed by a layer of liquid H_2. Densities of Al and liquid H_2 are 2.7 g/cm^3 and 0.07 g/cm^3, respectively. When the shock wave in Al reaches the interface with liquid H_2, the pressure in Al cannot be maintained by the much lighter and more compressible liquid H_2 and so the Al layer rapidly expands into the liquid H_2, which drives a shock wave into the liquid H_2 causing pressure in Al to release virtually isentropically.

Because P and u_p, and thus shock impedance $Z = P/u_p$, are continuous across an interface between two materials, the P/u_p state reached on isentropic release of shock pressure in Al is reached as well on the Hugoniot of liquid H_2 (Eq. 2.1). In Fig. 2.5, point B is on a curve that is the Hugoniot of Al reflected about a vertical line through point A. The reflected Hugoniot is a good approximation to the release isentrope, provided the release pressure is a significant fraction of shock pressure at point A. Point C illustrates small deviation of the two curves for Al under large shock-pressure release (Knudson et al., 2004).

The important point in Fig. 2.5 is that the release isentrope is essentially coincident with the mirror reflection of the Hugoniot at higher pressures; at lower release pressures the release isentrope deviates above the mirror reflection of the

Hugoniot. Deviation of the release isentrope from the mirror reflection occurs at point C, at which irreversible shock-compression energy becomes comparable to reversible PdV internal energy deposited by the shock at point A that remains in the material on release. Point C generally occurs at a release pressure of ~$(1/3)P_H$ (Altshuler et al., 1996). On total release of shock pressure, residual density $\rho_{res} < \rho_0$ because $T_{res} > T_0$. Residual temperature $(T_{res} - T_0)$ is transported out of a shocked material at late times by thermal diffusion.

At higher pressures, shock-compressed matter heats and cools in synchronization with hydrodynamic pressurization and release, respectively. A primary advantage of this process is that heat deposited hydrodynamically remains in the sample at high dynamic pressures during experimental lifetime. Thus, thermodynamic conditions calculated under shock compression are those in the sample during the brief time at extreme conditions. Slower cooling of residual irreversible shock heating is an issue for quenching high-pressure phases on release of dynamic pressure.

2.1.4 Shock-Dissipation Energy

By Eq. (2.3) total energy deposited by shock compression is represented by the area of the triangle under the line drawn from $(V_0, P_0,)$ to (V_H, P_H), which is called the Rayleigh line. Reversible energy is isentropic by definition and represented by the area under the isentrope in P-V space. Isentropic thermal pressure versus volume, $P_S(V,T)$, is generally small in the sense that isentropic pressure of a fluid on release nearly equals isothermal pressure at 0 K, $P_0(V)$, and so below ~100 GPa on the 0-K isotherm, generally $P_S(V) \approx P_0(V)$ (Nellis, 2005, 2006a).

Substantial irreversible energy E_d of Al shock-compressed to 150 GPa, for example, goes primarily into temperature and thus into thermal pressure $P(V,T)$ (McMahan, 1976). E_d is represented by the area between the Rayleigh line and the isentrope (Nellis, 2006a, p.1492). Entropy produced by shock compression of crystalline Al has little effect on pressure. Residual dissipation energy TS drives chemical reactions and phase transitions in shocked materials.

The split between dissipation energies T and S is determined by strength of interatomic interactions in the shock front. Relatively high compressibility of a van der Waals liquid means dissipation is generated primarily by a large volume collapse, which causes substantial shock heating T. For liquid Ar shock-compressed to 150 GPa, the area between the Rayleigh line and isentrope is about 90% of the area under the Rayleigh line (Nellis, 2006a).

Relatively low compressibility of strong crystals, such as for Al_2O_3, means substantial dissipation is generated by disordering the strong lattice. In this case, volume compression is resisted by strong interatomic bonds. Shock heating of the sample is relatively small because of the accompanying small volume decrease. In this case Hugoniot and isentrope are nearly equal. However, the Hugoniot of a strong solid, such as Al_2O_3, is not reversible in a microstructural sense because of shock-induced disorder of initially crystalline Al_2O_3 that remains in a sample on recovery at zero pressure, which release occurs on 100 ns times scales – too brief to anneal shock-induced defects generated at P_H.

2.1.5 Shock-Impedance Matching

Shock impedance is $Z = \rho_0 u_s = P/u_p$ by Eq. (2.1) with $u_0 = 0$ and $P_0 << P_H$. Thus $P_H = Zu_p$ for high shock pressure. As stated earlier, when a shock wave in one material crosses a boundary with a second material, pressure and particle velocity adjust so that P and u_p at the boundary are continuous and both materials move together with the same material velocity and pressure. *The shock-impedance matching principal states that shock impedance is matched across a boundary when a shock wave crosses a boundary between two materials.*

The simplest example of shock-impedance matching is a symmetric impact illustrated in Fig. 2.6. Impactor of Material I strikes Target also of Material I at impact velocity u_I. Since Impactor and Target are made of the same material with same initial density ρ_0, they both have the same R-H curve, which is a physical property of both. Impactor is initially travelling to the left at velocity u_I with $P = 0$. Target is initially at rest at $P = 0$. On impact, velocities of both along the vertical impact surface are $u_I/2$ by symmetry. Associated shock pressure on impact is also the same in both and given by their R-H curve. Along their interface Impactor instantaneously decelerates and Target instantaneously accelerates to $u_I/2$ both at impact shock pressure P_H. Impact shock pressure P_H is determined by the point of intersection of the right-going Hugoniot of Target with left-going Hugoniot of Impactor, as illustrated in Fig. 2.6.

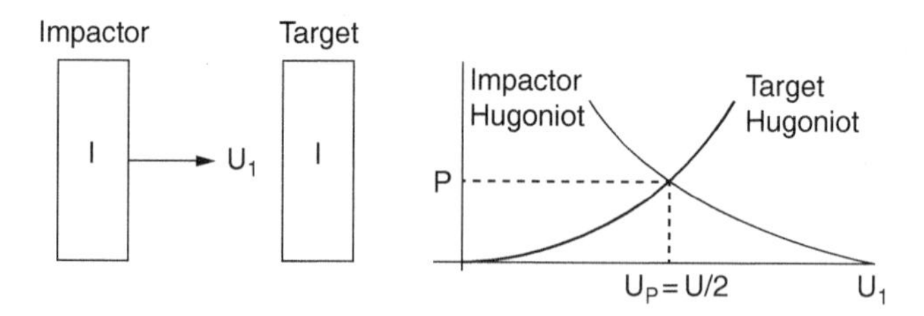

Fig. 2.6. Symmetric impact of Material I onto Material I to determine Hugoniot of Material I. Impactor and Target initially travel at velocities u_I and zero, respectively. After impact, both Impactor and Target travel to right at velocity $u_p = u_I/2$ at shock pressure P_H. u_I and u_s are measured in-flight and in Target, respectively. P_H, V_H and E_H are calculated with Eqs. 2.1 to 2.3.

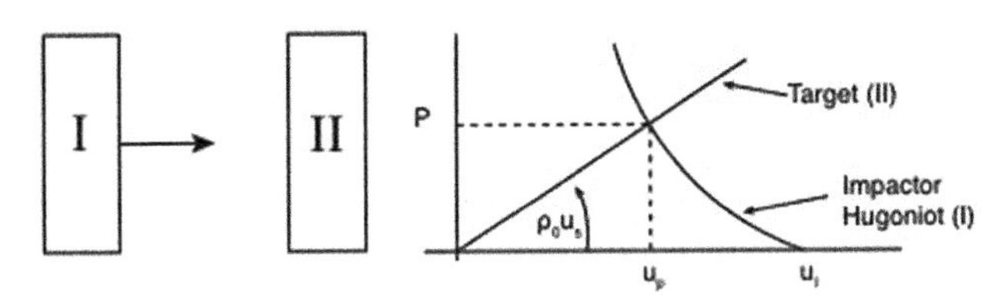

Fig. 2.7. Non-symmetric impact of Material I onto Material II to measure R-H curve of Material II once R-H curve of Material I is measured by symmetric impacts as illustrated in Fig. 2.6.

A non-symmetric impact is illustrated in Fig. 2.7. Once the R-H curve of Material I is measured with symmetric impacts in Fig. 2.6, Material I is used as Impactor to determined R-H curve of Material II used as Target. The R-H curve of Material I is plotted left-going from u_I and $P = 0$. u_I and u_s are measured. The state reached on impact occurs at the intersection of the left-going Hugoniot of Material I and Eq. 2.1 for Material II. The slope of the straight line in Fig. 2.7 is shock impedance of Material II, $Z = \rho_0 u_s$, which is also $Z = P_H/u_p$ on the Hugoniot of Impactor Material I. This experiment is another example of the shock-impedance matching method. Once u_s in Material II is measured and u_p is determined by shock impedance matching, then P_H, V_H and E_H of Material II are calculated with Eqs. 2.1 to 2.3. The process in Figs. 2.6 and 2.7 is repeated with several metals with a range of initial densities to obtain R-H curves that span a wide range of shock impedances and pressures.

Hugoniot pressure standards are generally given as C and S coefficients obtained from linear fits to $u_s(u_p)$ experimental data. Metals chosen to be qualified as shock pressure standards up to several 100 GPa include Al, Cu, Ta and Pt (Mitchell and Nellis, 1981b; Holmes et al., 1989; Nellis et al., 2003). Compendia of shock-wave data generated with chemical explosives have been published (Marsh, 1980; Trunin, 2001), some of which are used as pressure standards.

At higher shock pressures achieved with magnetic-flux-driven impacts, Al and SiO_2 quartz are used as pressure standards. The Hugoniot of Al and associated release adiabats have been measured up to 500 GPa (Knudson et al., 2003b, 2005). Similar measurements have been made for quartz up to 1600 GPa (Knudson and Desjarlais, 2009, 2013).

2.1.6 Error Bars of Shock-Compression Data

Experimental measurements require error bars in order for experimental data to be compared meaningfully with theoretical calculations. Fractional uncertainties in shock pressure $\delta P_H/P_H$ and shock compression $\eta = \rho/\rho_0$ are

$$\left(\frac{\delta P_H}{P_H}\right) = \left[\left(\frac{\delta u_s}{u_s}\right)^2 + \left(\frac{\delta u_p}{u_p}\right)^2_{\text{exp}}\right]^{0.5} + \left(\frac{\delta u_p}{u_p}\right)_{sys} \tag{2.7}$$

and

$$\left(\frac{\delta \eta}{\eta}\right) = (\eta - 1)\left[\left(\frac{\delta u_s}{u_s}\right)^2 + \left(\frac{\delta u_p}{u_p}\right)^2_{\text{exp}}\right]^{0.5} + (\eta - 1)\left(\frac{\delta u_p}{u_p}\right)_{sys}, \tag{2.8}$$

where $\delta u_s/u_s$ is the fractional uncertainty in measured u_s, $(\delta u_p/u_p)_{\text{exp}}$ is the fractional experimental uncertainty in determining u_p and $(\delta u_p/u_p)_{sys}$ is the fractional systematic uncertainty in u_p caused by systematic uncertainties in shock impedance matching. The experimental errors are random errors, which add as the square root of the sum of squares of individual experimental errors. Systematic errors are not random by definition and their absolute values add.

Shock impedance matching requires use of measured Hugoniots of pressure standards. A measured $u_s(u_p)$ Hugoniot is uncertain to the extent that measurement uncertainties of all the individual points used to determine a Hugoniot fit cause systematic uncertainty in the least-squares fit to that data, which is a function of u_p (Mitchell and Nellis, 1981b). Magnitudes of systematic errors add algebraically to random experimental error.

Fractional uncertainties are positive (absolute values). Because compression can reach values of ~4 at ~100 GPa shock pressures in fluids, uncertainties in η determined from Hugoniot experiments can reach up to ~thrice the uncertainty in measured u_s. Error propagation is substantial in shock experiments. It is important to measure u_s as accurately as possible to determine compression as accurately as possible for meaningful comparison with theory.

Double-shock experiments are those in which a sample is shocked twice. The initial state of the second shock is the first-shock state. Because the Hugoniot equations are used twice in the data analysis of double-shock data, error propagation can be quite substantial unless great care is taken. Uncertainties in the double-shock state are the convolution of uncertainties in determination of the first-shock state, experimental uncertainties in measurements of the double-shock states, plus systematic uncertainties in previously measured Hugoniots of all materials used in the data analysis (Nellis et al., 1981, 2003). Uncertainties in P_{H} and η in Eqs. (2.7) and (2.8) are caused by uncertainties in shock and particle velocities. Corresponding expressions for uncertainties in EOS need to be developed specifically for measurements made with ramp-wave compression as well. It is not meaningful to simply use formulas for uncertainties in Hugoniot data for uncertainties in data obtained with ramp waves.

2.1.7 Limiting Shock Compression and Quasi-Isentropic Compression of an Ideal Gas

The EOS of an ideal gas is useful for qualitative illustration of thermodynamic states achieved with dynamic compression. Shock compression produces very high temperatures relative to isothermal and isentropic compression. In fact, calculated temperatures and associated thermal pressures of an ideal gas become so large with increasing shock pressure that the R-H curve has a limiting compression η_L, beyond which no larger compression can be achieved no matter how large shock pressure is. That is, limiting compression $\eta = (\rho/\rho_0) \rightarrow \eta_L = (\rho/\rho_0)_L$ as $P,T \rightarrow \infty$. These properties are readily illustrated for a monatomic ideal gas with a limiting shock compression of 4-fold, $\eta_L = (\rho/\rho_0)_L = 4$. Fig. 2.8 is a comparison of

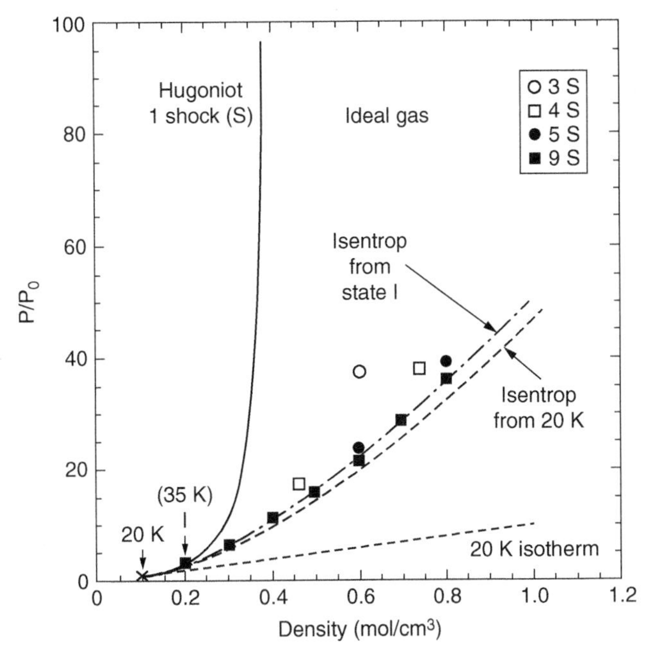

Fig. 2.8. Pressure versus density of ideal monatomic gas of H atoms initially at density 0.1 mol H/cm^3 and 20 K: Hugoniot (single-shock), multiple-shock curves from first-shock state H_1 at 0.2 mol H/cm^3, ideal-gas isentrope and isotherm starting from 0.1 mol H/cm^3 and 20 K. Calculations of multiple-shock compressions were performed for range of density from 0.2 to 1.0 mol H/cm^3 divided into 2, 3, 4 and 8 increments of equal density for a total of 3, 4, 5 and 9 shocks, respectively. As number of shocks increases, curve of P/P_0 versus density collapses from Hugoniot to isentrope from first-shock state at 0.2 mol H/cm^3 and 35 K. Calculated temperatures on Hugoniot and on isentrope from first shock state at 0.2 mol H/cm^3 and 35 K are 500 K and 100 K, respectively, at maximum pressures shown for each (Nellis, 2006a). Copyright 2006 by IOP Publishing. Reproduced with permission. All rights reserved.

calculated curves of adiabatic shock compression, isothermal static compression, isentropic compression and quasi-isentropic multiple-shock compression of an ideal gas.

The EOS of a monatomic ideal gas is $PV = (2/3)E$, $E = (3/2)RT$, and $\gamma = c_p/c_v = 5/3$, where c_p and c_v are heat capacities at constant pressure and constant volume, respectively. The initial state of the ideal gas of H atoms in Fig. 2.8 is $\rho_0 = 1/V_0 = 0.1$ g/cm^3 = 0.1 mol H/cm^3 at $T_0 = 20$ K and $P_0 = 17$ MPa. Limiting shock compression of an ideal monatomic gas is given by $\eta_L = (\rho/\rho_0)_L = (\gamma + 1)/(\gamma - 1) = 4$. The isotherm and isentrope of an ideal gas are given by $(P/P_0) = (\rho/\rho_0)$ and $(P/P_0) = (\rho/\rho_0)^\gamma$, respectively. The multiple-shock compression curve was calculated for an initial shock to 0.2 mol H/cm^3 and 35 K, followed by as many as eight shocks each of which was assumed to produce a density increment of 0.1 mol H/cm^3 from its preceding density up to a final density of 1.0 mol H/cm^3 at $\eta = 10$.

Fig. 2.8 illustrates differences between various curves of an ideal gas (Nellis, 2006a), which are qualitatively similar to differences between the corresponding curves of real materials. An important point in Fig. 2.8 is that shock compression of an ideal gas has a limiting compression of fourfold of initial density. On the Hugoniot at very high shock pressures and temperatures, matter becomes very hot and incompressible. Shock compression of gases is a way to make highly ionized plasma matter with degeneracy factor $T/T_F >> 1$, where T_F is Fermi temperature, at very high shock pressures.

An ideal gas of H atoms does not become metallic under single-shock compression because a limiting compression of fourfold is too small to make H a metal in degenerate condensed matter by overlap of $1s^1$ quantum mechanical wave functions on adjacent H atoms. However, historically a general conclusion drawn erroneously from this observation was that hydrogen cannot be metallized under any dynamic compression. While that conclusion is true for single-shock compression, it is *not* true for Q-I compression, say with multiple-shock compression.

Fig. 2.8 indicates that much higher compressions than fourfold at much lower temperatures than single-shock temperatures can be achieved under multiple-shock compression. This observation suggests the possibility that sufficiently high compression at sufficiently low temperatures might be achieved to make quantum-mechanically degenerate metallic H with multiple-shock compression. In fact, the simple calculation for the multiple-shock compression curve in **Fig. 2.8** was the basis of the multiple-shock experiments on liquid H_2 that made metallic fluid H with degeneracy factor $T/T_F \approx 0.014 << 1$. Because of its high density, MFH is cool condensed matter at ~3000 K, as are common metals at 300 K (Mott, 1936).

2.1.8 *Adiabaticity, Thermal and Mass Diffusion and Chemical Corrosion*

A supersonic shock wave travels in the longitudinal direction normal to the plane of the shock front. Thermal energy in condensed matter diffuses primarily by conduction at the speed of sound. Because shock compression is faster than thermal diffusion, shock compression is adiabatic. Specific internal shock energy is deposited only by shock compression itself. Of course, an experiment must be designed so that the experiment is in fact adiabatic.

In a finite-size sample in which shock propagation is initiated by an impactor with the shape of a short cylinder of radius R and sample thickness d or by a laser beam with a given radius, then it is possible in principle for shock-induced heat to diffuse radially outward from the shocked sample normal to the longitudinal direction of shock propagation. In this case, adiabaticity is insured in the portion of the sample probed diagnostically by choice of sample radius R to be large compared to sample thickness d ($R/d >> 1$), such that radial heat conduction out of the sample is prevented from the central volume of the sample for which measurements are made during experimental lifetime. That is, radial heat losses are effectively eliminated during experimental lifetime by choice of sample dimensions.

Mass diffusion coefficients at sonic velocities are comparable to thermal diffusion coefficients at sonic velocities. Thus, mass diffusion and thermal diffusion out of a sample are both generally negligible during experimental lifetimes. A useful exercise in this regard is to take the expression for diffusion from a point source and conservatively estimated diffusion constants and calculate diffusion distances during experimental lifetime. Negligible diffusion distance under shock compression is particularly important for dynamic compression of hydrogen. Light mobile hydrogen atoms heated under dynamic compression have insufficient time to diffuse out of their sample holder during experimental lifetimes of ~100 ns or less, as in 2SG experiments.

Chemical corrosion at interfaces between materials under dynamic compression is minimized by the fast timescales of shock experiments. For chemical corrosion to occur, interfacial diffusion followed by chemical reactions must occur. Both require time, and ~100 ns experimental lifetimes are generally too brief for significant corrosion to occur.

2.1.9 *Strength and Shock Propagation*

In a liquid or fluid, shock pressure is isotropic and generated by a step-increase in shock pressure (Fig. 2.2). In contrast, a solid has elastic strength (reversible) up to a pressure, or more properly up to a stress, called the Hugoniot elastic limit (HEL),

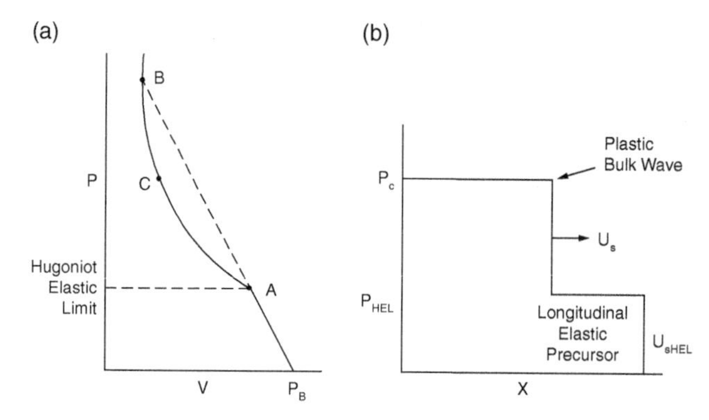

Fig. 2.9. Ideal elastic-plastic flow: (a) Hugoniot of solid with $P_{\mathrm{HEL}} = P_{\mathrm{A}}$. P_{B} is minimum shock pressure required to overdrive HEL, that is, to generate a single plastic shock wave. (b) At intermediate shock pressure P_{C}, two-wave structure propagates. P_{HEL} travels at longitudinal sound speed c_{L}, followed by slower plastic shock wave travelling at u_{s}, which increases with shock pressure for $P_{\mathrm{H}} > P_{\mathrm{HEL}}$. When u_{s} exceeds c_{L}, only a single plastic shock wave propagates and elastic precursor c_{L} is said to be overdriven.

point A in Fig. 2.9, above which a solid fails plastically (non-reversibly). In an ideal perfectly elastic-plastic solid the HEL is a constant independent of shock stress. Total stress is the sum of elastic and plastic stresses. For shock stresses somewhat greater than the HEL, a two-wave shock structure propagates because the velocity of an elastic shock wave at the HEL is greater than the velocity of a plastic shock wave just above the HEL. In an anisotropic (non-cubic) strong solid, wave propagation can be anisotropic depending on crystal structure.

For a solid with strength, stress and stress deviator are a more appropriate concept than pressure. This discussion is mainly about *ideal* elastic-plastic (e-p) shock flow, which occurs at stresses comparable to the HEL. At higher stresses well above the HEL, e-p shock flow is non-ideal and elastic strength might not be observed at all, for example, in strong solids at stresses above ~100 GPa. This discussion is meant as a caution that material strength might affect shock flows even at apparently very high shock pressures with expected fluid-like behavior.

Principal stresses σ_i in a solid and hydrostatic pressure P are given by

$$\sigma_i = -P + s_i, \qquad i = 1, 2, 3 \tag{2.9}$$

and

$$-P = (\sigma_1 + \sigma_2 + \sigma_3)/3, \tag{2.10}$$

where s_{i} are stress deviators in the principle directions i and $\Sigma s_i = 0$. For a solid with strength, P in Eqs. (2.1) to (2.3) is more properly replaced with normal stress in the direction of shock propagation, which is the longitudinal direction and one of the principal directions in a solid. In a strong cubic crystal, the other two principal directions, called the transverse directions, are orthogonal to each other and to the longitudinal direction.

The yield strength, or more properly the yield surface Y^0, is such that

$$s_1^2 + s_2^2 + s_2^3 \leq \left(Y^0\right)^2. \tag{2.11}$$

An ideal e-p solid is elastic up to a shock pressure or shock stress called the HEL, at which the solid fails plastically and is plastic at pressures greater than the HEL. The HEL of an ideal e-p solid is independent of shock pressure and constant in time, which is an idealization not always observed in real materials. Ideal e-p flow is illustrated in Fig. 2.9.

For pressure P_C such that $P_{HEL} < P_C < P_B$, a two-wave elastic-perfectly plastic shock-wave structure propagates. P_B is called the overdriven pressure, the lowest pressure at which only a single, plastic shock wave propagates. P_{HEL} of an ideal e-p solid is given by

$$P_{\text{HEL}} = \frac{c_L^2}{c_T^2}\frac{Y^0}{2}, \tag{2.12}$$

where c_{L}, c_{T} and Y^0 are longitudinal and transverse sound velocities, respectively, and Y^0 is yield strength (Kanel et al., 2004).

Shock-wave profiles in real materials with strength often differ substantially from ideal e-p flow, as illustrated in Fig. 2.10 for shock propagation in three

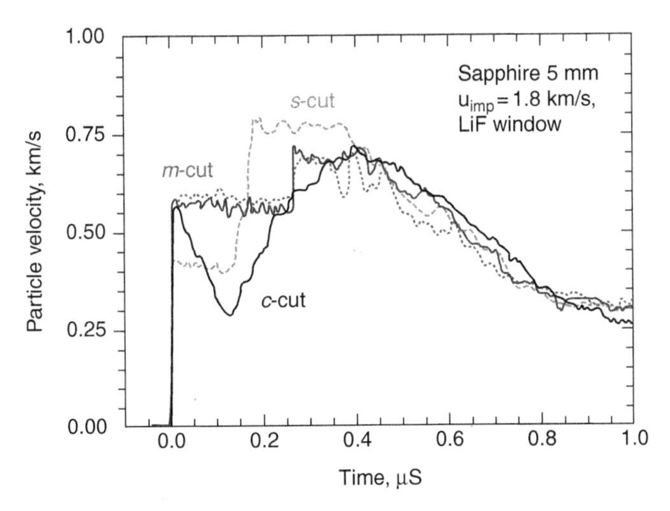

Fig. 2.10. Shock-wave temporal profiles measured for *m*-cut, *s*-cut and *c*-cut sapphire samples on 2 mm thick aluminum base plates impacted by 2 mm thick aluminum impactors at 1.8 km/s. Shock propagation is parallel and perpendicular to the *c* axis of *c*-cut and m-cut crystals, respectively. Shock propagation in *s*-cut crystals is in direction 72^0 from c axis (Kanel et al., 2009). Copyright 2009 by American Institute of Physics.

crystallographic directions of single-crystal Al_2O_3. Shock flows are anisotropic in the non-cubic rhombohedral crystal structure of Al_2O_3, as observed by shock-wave propagation in seven different crystallographic directions (Kanel et al., 2009). The HEL of *c*-cut sapphire is not constant in time and rapidly undergoes substantial elastic-stress relaxation prior to arrival of the plastic wave. Elastic strength of sapphire is not overdriven to "fluid" (single-wave) behavior until shock pressure exceeds ~90 GPa. Grady has reviewed shock flows in brittle materials (Grady, 1998). The HEL of GGG is pressure-dependent increasing from 7.65 to 24.5 GPa as final applied shock pressure increases from 8.52 to 89.5 GPa (Zhou et al., 2011).

At relatively high shock stresses and temperatures, ideal e-p flow is generally "smeared out" by dissipative effects. Strong single-crystal Al_2O_3 (sapphire) is used as anvils to multiply shock-compress H_2 above ~100 GPa to make MFH. The R-H curve of sapphire has been measured up to 340 GPa (Marsh, 1980; Erskine, 1994). At such high P_H and associated T_H flow of sapphire is fluid-like (single-wave) (Kanel et al., 2009).

Strong materials are interesting in themselves and because they are used as anvils for off-Hugoniot multiple-shock compression of low-density fluids. For these reasons it is important to experimentally characterize shock-wave propagation and to measure shock-induced electrical conductivities, optical transparencies and other physical properties of strong oxides used as anvils. Such measurements are needed in a number of strong materials with a variety of shock impedances to open up an extensive regime of ultrahigh pressures, densities and temperatures for experimental investigations of degenerate low-density materials under multiple-shock compression.

Velocity of the plastic shock increases with shock stress above the HEL. As plastic stress increases, velocity of the plastic shock increases until it eventually becomes comparable to the velocity of the HEL at a stress called σ_{OD}, the over-driven stress, which is point B in Fig. 2.9. At higher $\sigma > \sigma_{OD}$, only a single plastic shock propagates. Such a material can have elastic strength but it is not apparent as a two-wave shock structure, or wave profile, because the plastic shock travels at a velocity greater than that of the elastic one for $\sigma > \sigma_{OD}$. A single plastic shock wave is said to be indicative of fluid-like behavior, even if the material has not melted.

At extreme conditions considered herein shock pressures are generally sufficiently large that even solids with strong elastic strengths at ambient have fluid-like behavior in the solid phase for stresses σ in the range $\sigma_{OD} < \sigma < \sigma_{Melt}$, where σ_{Melt} or P_{Melt} is the stress or pressure on the Hugoniot at which that material melts on shock compression. For example, for GGG, σ_{OD} = ~110 GPa (Zhou et al., 2011) and σ_{Melt} = ~200 GPa (Zhou et al., 2015).

2.1.10 Thermal Equilibrium

If a shock-compression experiment at high pressure is repeated on a given material, experimental results generally repeat within the error bars. Such shock-compressed samples are said to be in mechanical equilibrium because such a measurement is reproducible. However, such shock-compressed samples are not necessarily in thermal equilibrium. Thermal equilibrium means a shock-compressed sample has a spatially uniform steady temperature in the region of the sample diagnosed during the time interval in which measurements are made. The R-H equations of themselves provide no information about thermal equilibrium.

The issue of thermal equilibrium must be addressed independently. For liquid Ar, whose atoms interact via effective van der Waals pair potentials, shock compression and heating thermally equilibrate rapidly in a narrow shock front, as indicated by results of molecular dynamics calculations in Fig. 2.2. An ensemble of independent particles in a fluid comes into thermal equilibrium by exchange of thermal energy in a large number of interatomic collisions. If a large number of interatomic collisions occur within the resolution time of an experimental diagnostic, then the ensemble of those atoms is probably in thermal equilibrium at the time a measurement is made.

The question of thermal equilibrium can be addressed by estimating the number of interatomic collisions within the resolution time t_{res} of a diagnostic system. This is an order-of-magnitude estimate. At high density, time t between interatomic collisions is estimated as $t = V_{atm}^{1/3}/v_{th}$, where V_{atm} is average volume per atom, $V_{atm}^{1/3}$ is average interatomic distance and v_{th} is average atomic thermal velocity of an ideal gas of atoms or molecules. Thermal velocity is estimated from calculated temperature T as $(1/2)mv_{th}^2 = (3/2)k_BT$, where m is atomic mass and k_B is Boltzmann's constant. If $t_{res} > \sim 10t$, then the fluid is probably in thermal equilibrium. In the case of MFH, $t \approx 10^{-14}$ s and $t_{res} \approx 10^{-9}$ s. Thus $t_{res} \approx 10^5\ t$ and MFH is in thermal equilibrium. Of course, thermal equilibrium could be addressed with molecular dynamics calculations.

Achieving thermal equilibrium at high densities under shock compression is generally not a problem for liquids and compressible solids. For example, the depth of the potential well between repulsive and attractive portions of the van der Waals pair potential of Ar is ~100 K. A well depth of ~100 K (0.01 eV) is small for temperatures above a few 100 K and P_H ~ 1 GPa.

In contrast, thermal equilibration of shock-compressed strong oxides is a more complex issue because strong oxides have covalent bonds with strengths of a few eV. Strong oxides are not expected to equilibrate thermally in bulk in a relatively thin shock front until shock pressures, compressions and thermal energy depositions essentially destroy those strong chemical bonds. For example, apparent shock-induced loss of the elastic strength of Al_2O_3 (sapphire) requires a shock pressure of ~90 GPa, which is the shock pressure above which only a single plastic shock is observed (Kanel et al., 2009), which means thermal equilibrium on shock compression of Al_2O_3 is not expected until shock pressure exceeds 90 GPa.

Calculations of pressures and temperatures of sapphire on its Hugoniot indicate sapphire melts and thus is probably equilibrated thermally at a shock pressure of 500 GPa at 10,000 K (Liu et al., 2015). Thus, thermal equilibrium under shock compression of Al_2O_3 probably occurs at a shock pressure, or stress, between 90 and 500 GPa. Exactly which shock pressure achieves thermal equilibrium rapidly on shock compression needs to be determined experimentally.

To date, shock pressure that achieves thermal equilibrium rapidly on shock compression of a strong oxide has been determined experimentally only for GGG, to the knowledge of this author. GGG equilibrates thermally in a thin shock front by scattering and absorption of photons at 130 GPa, at which the shock front of GGG is opaque (Zhou et al., 2015). Strong oxides equilibrate thermally at shock pressures of ~100 GPa, much greater than for compressible rare-gas fluids at ~1 GPa.

In strong solids at shock pressures below those comparable to dynamic strength, shock-compression energy is deposited heterogeneously. As a result the shocked crystal has essentially two distinct temperature regions. The single crystal breaks into an assembly of defected grains, whose interiors remain relatively cool, and are separated by grain boundaries and shear planes that are heated substantially and pinned at melting (Hare et al., 2002). Both of these shocked regions are probably in local thermal equilibrium but at substantially different temperatures. Because such a sample does not have a uniform temperature, the sample is not in bulk thermal equilibrium, even though most material in the two regions is probably thermally equilibrated locally.

In such a two-temperature shocked sample, a spatially unresolved spectral measurement is dominated by thermal emission from the region with highest temperature, which, for example, in Al_2O_3 is the grain-boundary/shear-plane region pinned at the melting temperature at pressure (Hare et al., 2002; Zhang et al., 2007). At sufficiently high shock pressures at which a crystal melts in bulk, uniform compression and shock temperature are expected.

2.1.11 Velocity of Sound

Release of dynamic high pressures occurs at the speed of sound. Pressure release occurs, for example, when a shock wave reaches a free surface at zero pressure or a when a shock wave crosses an interface from one material to a second material with lower shock impedance. An identical lower shock-release pressure is transmitted back into the first material with higher shock impedance. In contrast, a reshock occurs when a shock wave crosses an interface from one material to a second material with higher shock impedance. That is, the first shock reflects up in pressure when it reflects off an interface with a material that has higher shock impedance. An identical higher shock pressure is transmitted back into the first material with lower shock impedance. In both cases, P and u_p in both materials at the interface are continuous after a shock wave is transmitted in release or reflected up in pressure.

Pressure releases occur in timescales long compared to the rise time of a shock wave but generally too brief for significant thermal transport into or out of a material during pressure release. For this reason, pressure release curves are called release adiabats or release isentropes. Release isentropes are important for understanding lifetimes of a state achieved by dynamic compression, which is determined by edge releases at the speed of sound (Section 2.1.13). Velocities of sound are also important in generating shock waves that decay with time caused by a sound wave in shock-compressed material "catching up" to a leading shock.

Velocity of sound at a point on an R-H curve is produced by isentropic expansion from that point. Speed of sound c_{H} of a fluid on its Hugoniot is

$$c_H^2 = \left(\frac{\partial P}{\partial \rho}\right)_S. \tag{2.13}$$

Solids have relatively strong inter-atomic interaction potentials (such as covalent bonds), material strength and anisotropic physical properties, such as longitudinal and transverse velocities of sound. Strong solids do not necessarily equilibrate thermally under shock compression, depending on shock magnitude, inter-atomic bond strengths and experimental lifetime. Because a solid has strength it can support shear. Bulk sound velocity of a strong solid is given by

$$c_B^2 = c_L^2 - (4/3)c_T^2, \tag{2.14}$$

where c_B, c_L and c_T are bulk, longitudinal and transverse (shear) velocities of sound, respectively. For a fluid $c_T = 0$ and $c_B = c_L$. Shock pressure at which a solid melts can be determined in principle from Eq. 2.14 as the shock pressure at which c_T drops from a finite value to 0, which experimentally induces a discontinuity in c_B with P_H.

2.1.12 Decaying Shock Wave

Consider a shock front in a fluid that rises sharply to a steady pressure P_{H} for a while but eventually decays longitudinally because the energy source driving the shock cannot maintain a steady P_{H} indefinitely. Shock pressure decays from the

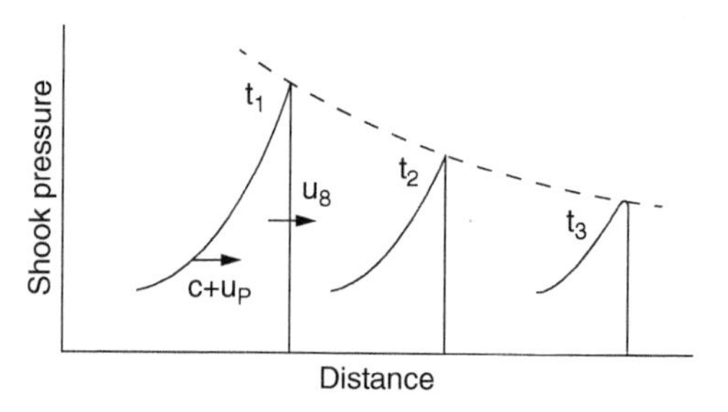

Fig. 2.11. Illustration of decaying shock wave caused by sound wave travelling at velocity c_H in medium travelling at particle velocity u_p catching up with leading shock front travelling at u_s. Hugoniot equations are obeyed across thin shock front and shock pressure reached at front is released initially at velocity of sound c_H. u_s, u_p and c_H decrease continuously with P_H and thus with distanced traveled.

rear of such a shock wave and the initial decay in pressure P_H occurs at c_H, the velocity of sound at P_H. Because shocked material moves at u_p, the pressure release wave is moving in the rest frame of the laboratory at velocity $u_p + c_H > u_s$. At ~100 GPa pressures, a rough rule of thumb is $c_H + u_p \sim 1.3\ u_s$. Thus, eventually the pressure release wave overtakes the leading shock wave. From that point on, P_H decays steadily with run distance and sound velocity decreases continuously as shock pressure releases. c_H can be determined by measuring travel distance in the sample at which the pressure release wave overtakes the leading shock front (McQueen et al., 1982).

A decaying shock wave illustrated in Fig. 2.11 is generated, for example, by a short (~few-ns) intense laser pulse or by a thin planar impactor onto a solid surface. The short pulse guarantees that a release wave rapidly overtakes the leading shock front. The hydrodynamics of a decaying shock have been derived (Trainor and Lee, 1982). Fig. 2.11 is a schematic of a series of spatial profiles of a decaying shock at increasing times t_1, t_2 and t_3. Doppler velocity history of the moving front and other fast continuous optical measurements can be made by reflecting a laser beam off the front of the shock wave as it propagates. This technique enables continuous measurements of Hugoniots and optical reflectivities, provided the sample equilibrates thermally in a time less than the resolution time of the optical detection system.

2.1.13 Phase Transitions: Graphite-Diamond, CaF_2 and $Gd_3Ga_5O_{12}$

Dynamic compression of condensed matter achieves high pressures, densities and temperatures, which means it is possible to drive a material across a phase boundary and induce a phase transition or crossover. The latter is also described as a continuous transition. Up to shock pressures of a few 10 GPa and shock

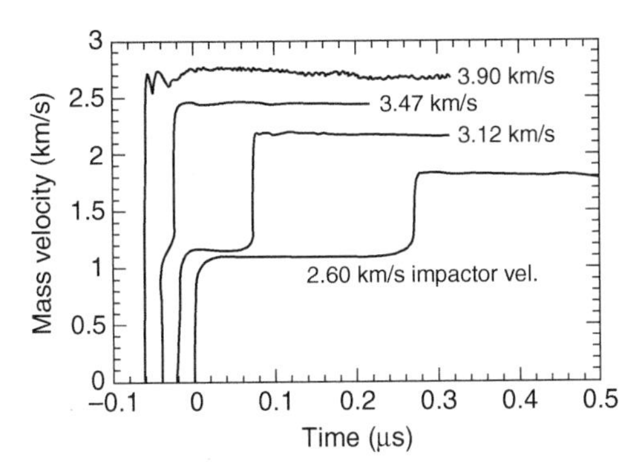

Fig. 2.12. Measured VISAR shock-wave profiles illustrating graphite-diamond transition for shock propagation parallel to *c* axis of graphite. First wave is elastic compression; second wave is phase transition (Erskine and Nellis, 1992). Copyright 1992 by American Institute of Physics.

temperatures of few 100 K, shock-induced phase transitions in solids can be quite sharp (Duvall and Graham, 1977).

A paradigm of a shock-induced phase transition is the graphite-diamond transition in highly oriented pyrolytic graphite (HOPG) at 20 GPa and ~500 K. Graphite sp^2 bonds are strong within basal planes and weak (van der Waals-like) in the *c* direction normal to basal planes of the graphite crystal structure. Measured wave profiles in Fig. 2.12 were generated by impacting HOPG *normal to basal planes* and parallel to the *c* axis using Cu plates accelerated with a 2SG to velocities in the range from 2.60 to 3.90 km/s. Mass velocity u_p was measured with a VISAR interferometer (Erskine and Nellis, 1991, 1992). The initial step increases near 1.1 km/s correspond to ~20 GPa at a relatively sharp graphite-diamond transition. Pressures corresponding to highest mass velocities in each experiment are 27, 35, 41 and 50 GPa, respectively. The graphite-diamond transition is essentially overdriven (complete) above ~50 GPa in the diamond or diamond-like high-pressure phase. Velocities of sound measured up to shock velocities of 18 GPa are sensitive to orientational disorder in HOPG (Lucas et al., 2015).

When HOPG is shock-compressed *parallel to basal planes* perpendicular to the *c* axis at shock pressures between 32 and 82 GPa, a somewhat different two-step process is induced (Sekine and Kobayashi, 1999; Sekine, 2000). Shock pressure of the first step occurs at a threshold pressure that depends on impact or final pressure of each experiment. In a sample compressed perpendicular to the *c* axis and recovered, scanning electron microscopy (SEM) observations of fracture surfaces show that, while basal planes are completely parallel in the un-shocked HOPG, platelet grains of recovered shocked samples are locally buckled and bent almost 90°. That is, in the second step, graphite platelets are re-oriented to be *perpendicular* to the original direction of shock propagation. In-plane bonding in HOPG is so strong that it is energetically favorable for platelets to re-orient as units, rather than decompose into amorphous or powdered C. Shock compression is an effective tool in studying effects of anisotropic strength of crystal lattices.

Shock pressures from 50 up to ~220 GPa have been obtained in diamond and lonsdaleite using a free-electron laser. In situ x-ray diffraction experiments were performed with ns x-ray pulses. Above 170 GPa, the direct transformation from pyrolytic graphite to lonsdaleite was observed (Kraus et al., 2016), a question that had been unresolved for decades.

While it is true that phase changes are generally observed as crossovers at 100 GPa shock pressures, there are two known exceptions. The R-H curve of compressible fluorite (CaF_2) has a sharp transition to a virtually incompressible phase at ~100 GPa shock pressure and associated high temperatures with compressibility comparable to that of diamond (Altshuler et al., 1973). This is noteworthy because CaF_2 is known as fluorite, a word derived from Latin that means "something that flows". In contrast, under static pressures of ~100 GPa, no such transition to a weakly compressible CaF_2 phase is observed (Dorfman et al., 2010). One possible explanation for the difference in CaF_2 under static and dynamic compression is shock-induced T and S, as in GGG at 120 GPa (Mashimo et al., 2006; Mao et al., 2011). Shock dissipation probably drives the sharp phase transitions at high temperatures at ~100 GPa in both CaF_2 and GGG.

2.1.14 Phase Crossovers: Liquid N_2 and CO

Above shock pressures ~15 GPa and temperatures of ~2000 K, phase transitions in fluids are generally broad and called crossovers or continuous transitions. Examples are dissociative and chemical-decomposition crossovers in the isoelectronic fluids N_2 and CO, respectively (Zubarev and Telegin, 1962; Nellis and Mitchell, 1980; Nellis et al., 1984b; Schott et al., 1985; Nellis et al., 1991a), illustrated in Fig. 2.13. These van der Waals liquids are isoelectronic and have the same density and normal boiling point at atmospheric pressure. The R-H curves of both are coincident up to ~15 GPa. Both liquids undergo molecular decomposition at shock pressures of ~20 GPa. N_2 probably dissociates to a fluid mixture of N and N_2; CO probably decomposes to a fluid mixture of C, O and CO. With further increases in shock pressure and temperature, both fluids approach limiting compressions of nearly fourfold of initial density. The implication is that above ~60 GPa shock pressure, N_2 becomes a fluid of N atoms, and CO probably becomes a fluid mixture of atomic C and O. At a much higher shock pressure, the Hugoniot of N_2 reaches a compression of 4.2-fold at 320 GPa (Trunin et al., 2008). One possible interpretation is that, above 60 GPa, shock temperatures are sufficiently large to ionize electrons from inner shells of the N atoms.

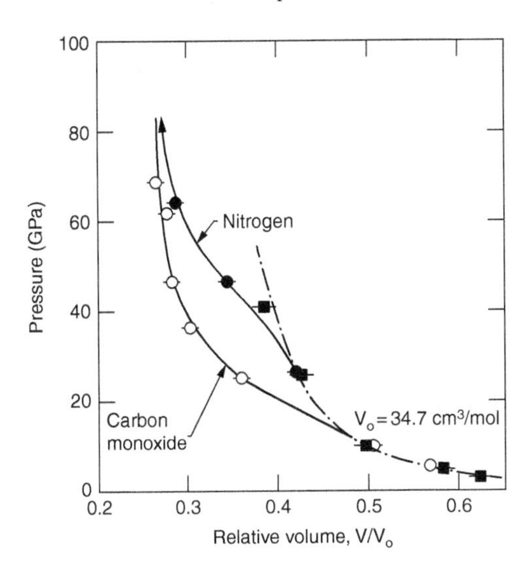

Fig. 2.13. Dissociative and chemical-decomposition crossovers above ~20 GPa in fluids N_2 and CO (Nellis et al., 1984b). Copyright 1984 by American Physical Society.

2.1.15 Common u_s (u_p) of Hugoniots of Liquid Diatomics

It is important to detect systematic behavior in dynamic compression data of various materials as an aid to understanding and as possible guidance to theoretical development. Fig. 2.13 indicates that P_H (V_H) of isoelectronic liquids N_2 and CO are identical for $V/V_0 > 0.5$ and $V/V_0 < \sim 0.3$. At the larger relative volumes these liquids are diatomic; at the smaller relative volumes these liquids are probably monatomic N and C/O. Initial densities ρ_0 of liquid N_2 and CO are identical at initial (atmospheric) pressures in those experiments.

The observation in Fig. 2.13 raises the question as to the relative systematic behavior of u_s (u_p) of several diatomic liquids N_2, CO, O_2 and H_2/D_2, which have various values of ρ_0. Fig. 2.14 shows that all four diatomic liquids have common systematic behavior on their Hugoniots in u_s (u_p) space. Slope S of the diatomic H_2/D_2 liquids is 1.22 (Nellis et al., 1983), as well as for the liquids N_2, CO and O_2. However, common systematics of the Hugoniots of these liquids are not observed generally in P_H (V_H) space (Nellis, 2002b). These results suggest that a systematically similar behavior of fluids in general is more likely to be found in u_s (u_p) than in P_H (V_H) space. Because all these fluids interact by effective van der Waals pair potentials independent of initial densities (Ross and Ree, 1980) and all these diatomic liquids have Hugoniot slope S = 1.22, their appears to be a direct uniform relation between slope S and effective van der Waals pair potentials, independent

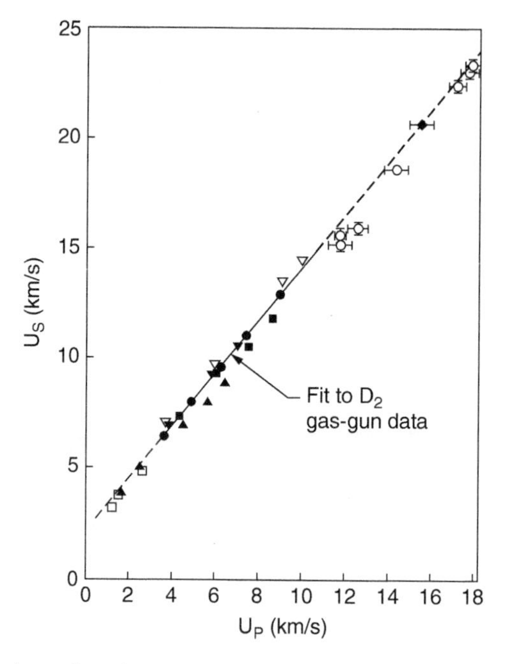

Fig. 2.14. Hugoniots plotted as u_s versus u_p for liquid D_2, H_2, N_2, CO and O_2. Shock compression was generated with two-stage gas gun and with Z Accelerator for the D_2 data at highest velocities. u_s (u_p) of diatomic phases are consistent with linear fit to D_2 data shown. Dissociation of D_2, N_2 and CO is observed as ~3% decreases in u_s with respect to common fit for diatomic phase (Nellis, 2002a). Copyright 2002 by American Physical Society.

of liquid densities at ambient pressure and potential parameters that scale with critical points of the various liquids.

2.1.16 Computer Simulations of Longitudinal Planar Shock Propagation

Impactors, laser beams and sample holders are finite in size. For this reason a crucial aspect of dynamic-compression experiments is that sample holders be designed such that the volume of compressed sample diagnosed is free of longitudinal and radial edge effects during experimental lifetimes. In practice this means that, if the leading shock front is diagnosed, then that leading shock front is unaffected by longitudinal or radial edge effects. That is, sample volumes diagnosed are effectively compressed in planar geometry as far as the diagnostics can detect. Shock compression experiments are generally designed with 1D, planar, hydrodynamic computational simulations of shock propagation. Such simulations are used to design a uniform (P, V, T) state in a material during the time interval in which measurements are made.

This goal is accomplished by choosing various layer thicknesses such that unwanted longitudinal shock-wave reflections and releases in a specimen do not affect a sample volume in a uniform thermodynamic state for diagnostic measurements during experimental lifetime. With respect to Fig. 2.1, for example, a logical question to ask is, what are the thicknesses of impactor, the left wall of the liquid Ar layer and the liquid Ar layer itself such that hydrodynamic affects originating from the impact surface from the rear (left) surface of the impactor and from the front (right) surface of the wall with Ar do not affect diagnostic measurements during experimental lifetime?

For 1D planar-shock flow of a fluid, the equation of mass motion is (Wilkins, 1999)

$$\rho \frac{\partial u}{\partial t} = -\frac{\partial}{\partial x}(P + Q), \tag{2.15}$$

where u is mass velocity and Q is von Neumann artificial viscosity, a mathematical construct to track shock-front discontinuities in a finite-element hydrodynamic simulation code (Von Neumann and Richtmyer, 1950). Q is proportional to some power of $(\partial u/\partial x)$because the position of a shock front corresponds to where the gradient of mass velocity is large. Q is essentially zero except at the travelling shock front.

The equation of mass continuity is

$$\frac{\partial V}{\partial t} = V \frac{\partial u}{\partial x}. \tag{2.16}$$

The energy equation is

$$\frac{\partial E}{\partial t} = -(P + Q)\frac{\partial V}{\partial t}. \tag{2.17}$$

In a finite difference solution, initial values of $P,\ V, E$ and Q are used to calculate changes in u (Eq. 2.15), V (Eq. 2.16) and E (Eq. 2.17). The new values of V and $E$are used to calculate new P with an equation of state, and the new u is used to calculate the new Q. The new values are cycled for iteration. Various combinations of layer thicknesses and materials can then be tried to search for an optimal design of a dynamic-compression experiment. For example, calculated planar shock-wave structures for double-shocked liquid N_2 experiments are published (Nellis et al., 1991a).

2.1.17 Radial Edge Effects

In 1D adiabatic shock experiments, uniform shock pressure, density and temperature are required in a macroscopic volume in a sufficiently long time interval to

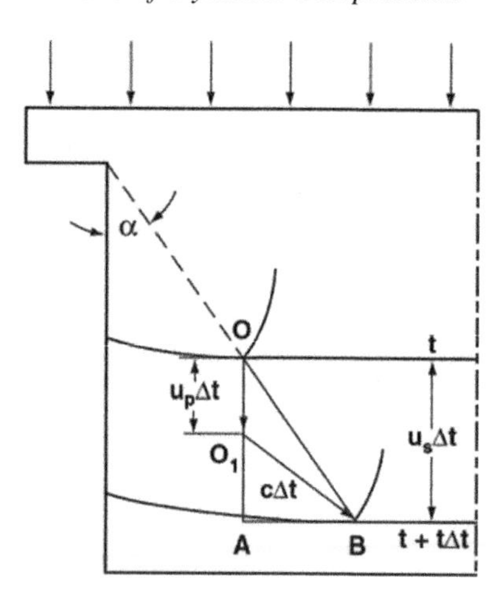

Fig. 2.15. Illustration of radial edge release of incident shock pressure in time interval Δt in direction transverse to direction of shock propagation, which is vertically downward. Shock velocity is u_s, particle velocity is u_p and c is sound speed at incident shock pressure. Angle α is angle of radial pressure release (Altshuler et al., 1960).

make accurate measurements of desired quantities. Diagnostic resolution time must be much less than experimental lifetime, which is determined by radial edge releases and longitudinal wave reflections at interfaces in a sample holder. In planar experiments, measurements are usually made on adiabatically compressed material (1) near the center of a cylindrical sample with sample thickness much less than sample diameter to avoid radial edge-release effects, and (2) thicknesses of layers in the sample holder are chosen to avoid unwanted longitudinal wave reverberations reaching the leading shock front during experimental lifetime. By so doing, a sample at extreme conditions can be diagnosed in a 1D geometry before edge effects arrive and destroy the 1D symmetry in the region diagnosed. A sample holder designed to be safe against hydrodynamic edge effects is also generally safe against thermal-conduction edge effects, because both travel at sound speed. Layer thicknesses chosen to avoid longitudinal wave reflections can be calculated manually with the method in McQueen et al. (1982) or with hydrodynamic computational simulations described above.

Fig. 2.15 shows a planar shock front incident from above onto a flat circular surface. As the planar shock wave propagates longitudinally downward at u_s, it encounters a corner from which the shock-compressed cylinder expands radially outward at the velocity of sound of the incident shock pressure. Fig. 2.15 also illustrates the extent to which the central circular region continues to travel un-rarefied downward at initial shock velocity u_s with particle velocity u_p. The radial pressure release wave travels inward at angle α subtended by the initial outer radial boundary of the sample at the perturbing corner and the trajectory of the outer radius of the un-rarefied volume at time t, with t = 0 at shock arrival time at the corner. Altshuler et al. (1960) have shown that

$$\tan\alpha = \left[c^2 - \left(u_s - u_p\right)^2\right]^{1/2} / u_s. \tag{2.18}$$

Experiments are designed so that experimental diagnostics probe an area conservatively less than a radius defined by angle α at termination of the experiment, as given by Eq. 2.18.

2.1.18 General Materials Effects

Shock compression produces a number of effects on materials, including shock compaction of powders (Benson and Nellis, 1994), shocked ejecta from natural impacts (Gratz et al., 1993a), shock-induced amorphization (Gratz et al., 1993b) and potential synthesis of novel superconductors (Nellis et al., 1988b). Application of static pressure enhances alloy solubility (Dubrovinskaia et al., 2005) and might do so under dynamic compression as well. Because melting of dense oxides at high pressures and temperatures induces electrical conductivities (Liu et al., 2015), fluid oxides at high P/T deep in exoplanets might make contributions to magnetic fields of those bodies by convection of electrically conducting fluid oxides (Nellis, 2012).

2.2 Quasi-Isentropic Multiple-Shock Compression

The goal of our first quasi-isentropic (QI) experiments with this technique (Weir et al., 1996a) was specifically to make a metallic phase of dense monatomic hydrogen using dynamic compression. At the time of those experiments, metallic solid hydrogen had yet to be made under static compression alone by $E = -\int P\, dV$at ~300 K. At finite T phase stability is controlled by Helmholtz Free Energy $F = E - TS$. Temperature T = 300 K is negligible compared to the binding energy of the H_2 molecule, 4.5 eV. So essentially at T = 300 K F of an H_2 molecular system under static compression is $F \approx E$. Given that the search for metallic solid hydrogen began in earnest around 1980, by the 1990s it was time to employ the TS term in F by generating T and S with dynamic compression.

WH (1935) predicted a metallic H density of 0.62 mol H/cm^3 at "very low temperatures". Based on estimates in Fig. 2.8 using the ideal gas EOS, limiting shock compression of liquid hydrogen starting from 0.07 g/cm^3 is estimated to be about fourfold (0.28 g/cm^3 = 0.28 mol H/cm^3) at very high temperatures. On the other hand, multiple-shock compression of an ideal gas is coincident with the isentrope from a first-shock state up to tenfold compression at $T_{MS} \approx T_H/5$, where T_{MS} and T_H are the multiple-shocked and shock temperatures, respectively. The ideal gas equation of state was used as a model EOS to identify a likely compression history required to make metallic H. Those considerations motivated our attempt to make MFH with multiple-shock compression.

2.2.1 MFH by Quasi-Isentropic Compression

Single-shock compression of liquid H_2 produces too much temperature and insufficient density to make MFH by overlap of $1s^1$ wave functions on adjacent H atoms. To test WH's prediction for condensed matter, it is essential to make an H sample of condensed matter – that is, a degenerate electronic system. WH's predicted density of dissociation and metallization is 0.62 mol H/cm^3 = 0.31 mol H_2/cm^3 (Wigner and Huntington, 1935), which occurs at 73 GPa and 300 K (Loubeyre et al., 1996). This density is ninefold H density in liquid H_2 at 20 K.

In the 1990s, compression of solid H_2 at ~300 K was yet to be observed to dissociate H_2 to H and make metallic solid H (MSH) at static pressures above ~300 GPa, which suggests insufficient $\int PdV$ work has been deposited in solid hydrogen to drive dissociation and metallization by pressure alone. Sufficient heating of solid H_2 in a DAC is expected to drive dissociation of H_2. However, heating hydrogen in a DAC at static high pressure is problematical because hydrogen diffuses out of its sample holder within a few ms (McWilliams et al., 2016). Dynamic experiments with lifetimes significantly less than ms were needed.

Dynamic compression generates P, T and S simultaneous with compression ρ. Thus, a logical way to try to make metallic H is to use dynamic compression to generate T and S, which might dissociate H_2 to H at T thus creating S and perhaps change the stable phase from H_2 to H. Cryogenic capability existed at LLNL to make liquid H_2 samples since the early 1980s. Fast electronic diagnostics to measure electrical conductivity with ~ns time resolution in experiments with 100 ns lifetimes had been in hand since the ~1970s. Thus, the possibility to make and diagnose metallic hydrogen on a fast timescale under dynamic compression was in our hands.

To maximize the probability of making metallic H at ~0.62 mol H/cm^3 the highest pressures and densities at lowest temperatures possible were achieved by dynamically compressing liquid H_2 quasi-isentropically by multiple-shock compression. A sequence of ~10 relatively weak shocks up to a peak pressure of ~100 GPa was achieved by reverberating a shock in liquid H_2 contained between two weakly compressible oxide anvils. In the 1990s the only incompressible electrically insulating material available to us as anvils in our dynamic-compression experiments was sapphire, single-crystal Al_2O_3.

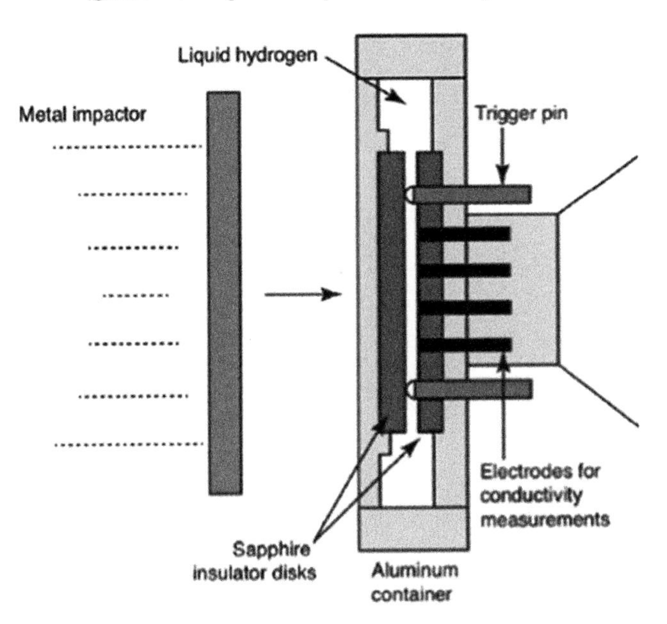

Fig. 2.16. Dynamic compression experiment on liquid H_2 initially at 20 K. Liquid sample diameter and thickness are 25 and 0.5 mm, respectively. Impact generates shock in liquid H_2 that reverberates more than nine times between sapphire anvils (Nellis et al., 1999). Copyright 1999 by American Physical Society.

Our goal was to dissociate H_2 to H at sufficiently high ρ and low T to make degenerate metal H ($T/T_F<<1$), where T_F is Fermi T of the Fermi-Dirac distribution. Melting temperatures of H_2 at 100 GPa pressures were then yet to be measured. However, because substantial dissipation in the form of T and S would be generated at 100 GPa shock pressures, sample Ts would probably be greater than melting temperatures T_m. Thus, the dynamic-compression process might make dense fluid H that hopefully would be metallic (MFH).

Quasi-isentropic off-Hugoniot states in liquid H_2 are readily achieved by (1) tuning the shape of the pressure pulse applied to a sample and by (2) a slight revision of the configuration of a single-shock cryogenic sample holder at 20 K developed in the early 1980s (Nellis et al., 1983). For multiple-shock experiments, the liquid H_2 samples were disks with ~25 mm diameter and ~0.5 mm thickness contained between two high-density electrically insulating sapphire anvils each with thickness ~2 mm. A classical two-stage light-gas gun (Jones et al., 1966; Isbell, 2005) generated pressures up to 180 GPa by impact. Multiple-shock compression was obtained in situ with a pressure pulse in liquid H_2 as in Fig. 2.16 (Nellis et al., 1999).

Because WH predicted an insulator-metal transition (IMT) from H_2 to H at ninefold compressed atomic H density in liquid H_2, many step compressions are

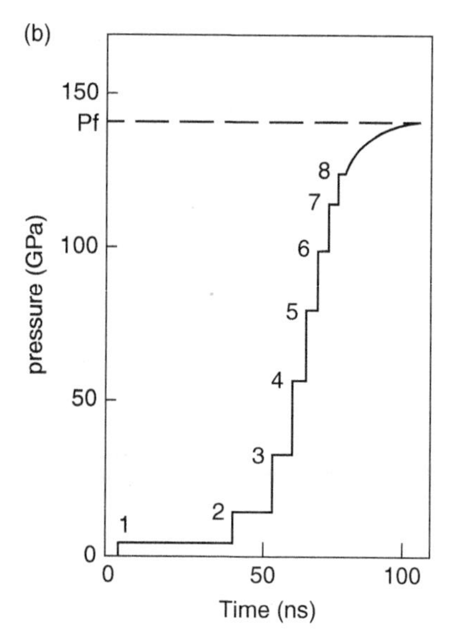

Fig. 2.17. Shock pressure versus time in middle of hydrogen layer in Fig. 2.16 calculated with one-dimensional hydrodynamics code. Numbers on the steps are first-shock, second-shock, third-shock, and so forth of wave reverberating between sapphire disks (Nellis et al., 1999). Copyright 1999 by American Physical Society.

needed to achieve such a large compression. For conductivity experiments the anvils (1) need to be electrical insulators, (2) have large shock impedance, (3) have Hugoniot and release isentropes that are nearly identical (reversible) for simplified computational design and (4) the Hugoniot of the anvils must be known to design the sample holder. Sapphire satisfies these requirements. Shock dissipation in sapphire appears primarily as disorder (S), rather than T and thermal pressure. The density of sapphire is 4.0 g/cm^3, sixty times greater than the density of liquid H_2, 0.07 g/cm^3 at atmospheric pressure. The large difference in shock impedance means many shock reverberations are achieved to quasi-isentropically compress liquid H_2.

Shock hydrodynamics of wave reverberation in the (Al_2O_3-liquid H_2-Al_2O_3) "sandwich" at the center of the sample holder in Fig. 2.16 are shown as P versus time in Fig. 2.17, which is a plot of calculated pressure history at the midpoint of the hydrogen layer. Numbers on the steps in Fig. 2.17 indicate the first shock, second shock and so forth of the reverberations travelling back and forth between the two sapphire disks in Fig. 2.16. Fig. 2.17 also shows time durations of the steps. Shock velocity increases with shock pressure and so the duration of each step decreases as shock pressure and velocity increase and sample thickness decreases during the compression process. Electrical conductivities were measured at maximum "ring-up" pressures in the range between 90 and 180 GPa. Peak calculated temperature in each of those experiments range from 1500 to 2900 K. In this range of finite temperatures, dense fluid hydrogen crosses over from semiconductor to degenerate poor metal ($T/T_F \approx 0.014$) at 140 GPa at the density predicted by WH (1935). At metallization density in those experiments, calculated $T = 2600$ K and at metallization density ρ_{met}, $T_F = 220{,}000$ K.

2.2.2 *Interfacial Instabilities: Rayleigh-Taylor and Richtmyer-Meshkov*

A concern about the sample holder in Fig. 2.16 is potential hydrodynamic instability and a resulting mixing of sapphire and liquid H_2 along their interface, which might be caused when a ~100 GPa (Mbar) shock wave in sapphire with density 4.0 g/cm^3 breaks out into liquid H_2 with density 0.07 g/cm^3. In this case surface imperfections in interfacial sapphire, inspected to be flat to 30 nm peak-to-peak, might grow and induce turbulent interfacial mixing, which might substantially degrade a conductivity measurement of dense hydrogen. To minimize probability of such mixing in Fig. 2.16, several steps were taken, as will be discussed.

J. W. Strutt, Lord Rayleigh, (1883) and G. I. Taylor (1950) developed theories of R-T instability growth along the interface between two fluids with different densities under constant acceleration. The R-M instability occurs at the interface between two fluids that are loaded impulsively by a shock wave (Mikaelian, 1985). R-T and R-M theories are used to assess the likelihood that such interfaces are hydrodynamically stable (Barnes et al., 1974).

For the shock-induced R-M instability, the lowest order growth rate of a periodic interfacial perturbation is

$$\eta(\tau)/\eta(0) = 1 + \gamma\tau, \quad \gamma = \Delta\varpi\kappa A, \tag{2.16}$$

where $\eta(\tau)$ and $\eta(0)$ are amplitudes of a sinusoidal perturbation at time τ and $\tau = 0$, respectively, λ is the wavelength of the perturbation along the interface, $k = 2\pi/\lambda$, ρ_1 and ρ_2 are the densities of the two fluids, Δv is the jump velocity caused by passage of a shock wave from ρ_1 to ρ_2 and A = $(\rho_2 - \rho_1)/(\rho_2 + \rho_1)$ is the Atwood number. Because 4.0 g/cm^3 >> 0.07 g/cm^3, A and thus growth rate γ are near maximum at the sapphire/hydrogen interface in Fig. 2.16. However, sapphire is one of the strongest materials known and thus expected to resist instability growth substantially more strongly than fluids and solids that are much more compressible. The sapphire disks were optically flat with surface perturbations of ~30 nm determined by mechanical scanning of the sapphire surfaces. Experimental lifetimes were ~100 ns, extremely brief, which minimizes possible growth time of interfacial perturbations.

2.2.3 *Minimizing Interfacial ICF Instabilities: Multiple-Shock Compression*

In ICF experiments, pressures, temperatures and velocities are many orders of magnitude larger than in shock reverberation experiments in Fig. 2.16. Concern about R-M instabilities is much larger for ICF, which is a program to generate an alternative source of commercial energy by laser-driven nuclear fusion of D-T (Motz, 1979; Lindl et al., 2014; Hu et al., 2015). Interfacial instabilities are an issue in astrophysics as well (Calder et al., 2002). In ICF experiments at the National Ignition Facility (NIF) at LLNL, multiple pulsed laser beams drive dynamic compression in a fuel pellet consisting of an ablator surrounding molecular D-T (D_2-T_2) fuel. Laser pulses generate multiple-shock waves in the ablator, which in turn compress D-T fuel to high dynamic pressures, densities and temperatures to generate thermonuclear D-T burn in the center of a spherical capsule.

A problem arises because fusion burn rate R of D-T is $R \propto \rho^2 T^4$, where T is temperature (not tritium). If particles from the ablator are mixed with dense fluid D-T heated by dynamic compression to a few 100 eV, those particles are ionized by absorbing thermal energy from dynamically compressed fluid D-T, which reduces temperature of the gas, lowering burn rate as the 4th power of T – a very large effect. Because such deleterious effects might be so large, it is crucial to minimize possible mixing of ablator and D-T fuel.

To reduce ablation-front-driven R-M instability growth and subsequent degradation of fusion-energy yield, the shape of incident laser pulses is tuned to generate a sequence of multiple-shock waves that generate gentler, smoother accelerations across material interfaces, which are expected to reduce interfacial mixing and instability growth to produce higher fusion-energy yields than had been achieved previously with fewer but stronger shock waves. Dynamic QI compression developed to make metallic fluid H appears also to contribute to solving the problem of the R-M instability (Hurricane et al., 2014). The latter point emphasizes potential benefits to be gained by integrating basic scientific research with applied research and development.

3
Generation of Dynamic Pressures

Dynamic pressures generated experimentally have ranged from $\sim 10^{-2}$ to $\sim 10^{5}$ GPa (100 GPa = 10^{6} bar = 10^{-1} TPa) starting in the 1940s. Their purpose has been to measure material properties at extreme conditions. Dynamic-compression facilities generate pulsed high pressures by rapid energy deposition. From the 1940s to 1960s dynamic compression was obtained with shock waves generated by chemical explosives in contact with specimens. Beginning in the 1950s, shock drivers have used rapid deposition of kinetic energy generated with chemical explosives and guns that accelerate projectiles to high velocities prior to impact. These shock drivers include single-stage gas guns, two-stage light-gas guns and mass accelerators driven by pulsed magnetic pressures. Shock waves generated with planar or hemispherical chemical-explosive systems accelerate metal impactors to velocities up to ~5 km/s and ~15 km/s, respectively (Rice et al., 1958; McQueen et al., 1970; Marsh, 1980; Altshuler, 1965, 2001; Zhernokhletov, 2005).

From the 1960s to ~1990 dynamic pressures up to $\sim 10^{5}$ GPa range were generated in proximity to underground nuclear explosions (Trunin, 1998 and 2001; Ragan, 1984; Mitchell et al., 1991). In recent years dynamic pressures in excess of 10 TPa have been generated with high-power pulsed lasers, such as the NIF at LLNL (Bradley et al., 2004). Pressures and densities achieved in all the above dynamic experiments were calculated with the R-H equations; associated temperatures must be calculated for the vast majority of those cases.

Sample dimensions and experimental lifetimes depend on the areal-energy density of the impactor or of the radiation beam used to generate dynamic pressure. Sample size and experimental lifetime are limited by radial and longitudinal pressure release and/ or re-shock edge effects. In the cases discussed herein, samples generally have sufficiently large dimensions, and experimental lifetimes are sufficiently long to measure material properties with available time resolution with accuracies of a few percent.

In the 1960s dynamic high pressures began to be generated with gas dynamics in a two-stage light-gas gun (2SG) in which compressed H_2 gas accelerates a 20 g metal impactor plate to velocities as high as ~8 km/s (Charters et al., 1957; Jones et al., 1966; Mitchell and Nellis, 1981a; Nellis, 2000, 2007a). The 20 m long 2SG was developed for materials research on a laboratory scale. Impact velocity u_I is measured in free flight, and shock velocity is measured in the target. This method has the advantage that Hugoniots of solids can be measured purely by the experimental methods illustrated in Figs. 2.6 and 2.7. Hugoniots of Al, Cu, Ta and Pt have been qualified in 2SG experiments as EOS standards up to shock pressures of a few 100 GPa (Marsh, 1980; Mitchell and Nellis, 1981b; Holmes et al., 1989; Trunin, 2001). Measured shock pressures typically range from 10 to 500 GPa for liquid H_2 and Ta, respectively.

Impact velocities u_I up to ~15 km/s and shock pressures greater than ~TPa have been achieved with hemispherically convergent, chemical-explosive systems (Altshuler et al., 1996; Trunin, 1998). H_2 gas in a linear 2SG and gaseous reaction products of chemical explosives in a hemispherically convergent system achieve maximum impactor velocities limited by the speed of sound in the dense molecular gas pushing the impactor. To obtain still higher impact velocities, a driving gas is needed that is limited by a speed much greater than the speed of sound in a molecular gas.

More recently impact velocities as large as ~45 km/s have been achieved with the Z Accelerator at Sandia National Laboratories Albuquerque (SNLA) (Schwarzschild, 2003). Those velocities are driven by magnetic pressure generated by fast, magnetic-flux compression (Hall et al., 2001; Lemke et al., 2011; Knudson and Desjarlais, 2013). The Z Accelerator is an analogue of a 2SG in the sense that the compressed gas is a "gas" of magnetic flux rather than H_2 molecules. Magnetic flux moves at the speed of light. The history of dynamic compression at SNLA is documented (Asay et al., 2017).

The goal that motivated construction of the large NIF facility at LLNL (Bradley et al., 2004) and the Z facility at SNLA is achievement of ICF, a potential practical source of commercial power. Development of those two techniques has resulted in a substantial facility size. To achieve shock pressures above ~TPa, it is essential to deposit sufficient energy over a sufficiently large sample over a sufficiently long pulse duration to make accurate measurements. For this reason giant pulsed lasers and pulsed-power sources that achieve TPa pressures at U.S. national laboratories are enormous compared to, say a 20 m long 2SG that generates pressures up to a few 0.1 TPa for materials research purposes. The NIF pulsed laser facility occupies a volume comparable to that approximately of a cubic football field.

To achieve ICF, it is essential to know properties of materials at relevant extreme conditions. Thus, it is important to perform scientific experiments to measure properties of materials needed to design experiments to achieve ICF. Historically, extensive dynamic compression experiments have been performed up to ~100 GPa using shock pressures generated with plane-wave chemical explosives and single-stage gas and gunpowder guns. In this chapter we describe experiments that drive dynamic compression experiments well above 100 GPa pressures.

To put kinetic energies of shock-wave drivers into perspective, kinetic energy of a 20 g impactor accelerated to 7 km/s with a 2SG is ~0.5 MJ, which is comparable to total kinetic energy of the 10^{12} protons and antiprotons in colliding beams at the Tevatron at the Fermi National Accelerator Laboratory (Fermilab). Kinetic energy of impactors generated by fast planar magnetic-flux compression at the Z Accelerator is substantially greater, comparable to energies in the particle beams in the Large Hadron Collider (LHC) at CERN. Energies of the NIF Laser and Z mass Accelerator are used to study novel states of atomic matter (ultracondensed matter, warm dense matter and dense plasmas) analogous to the way high-energy particle beams at the LHC are used to probe novel states of sub-nuclear matter.

3.1 Two-Stage Light-Gas Gun

The 2SG was developed in the 1950s and 1960s to perform research on Earth that would enable manned space flight to the moon by the end of 1960s, as called for by U. S. President John F. Kennedy. The primary concern was the safety of the crew in spacecraft on their voyage through space. The threat is impact of micro-meteorite particles travelling at impact velocities greater than 15 km/s. Half that velocity was readily obtainable on Earth to assess the likely magnitude of damage that would be caused by such impacts. LLNL obtained a 2SG to perform experiments on materials at extreme conditions of P and T.

The 2SG at LLNL, illustrated in Fig. 3.1, accelerates an impactor to velocities as high as 8 km/s using compressed H_2 gas. The gun breech contains up to 3.5 kg of gunpowder and the pump tube is filled with ~60 g of H_2 gas initially at ~0.1 MPa pressures. Depending on impactor material, target material and impact velocity, dynamic pressures in a target can be tuned from ~GPa to as much as ~650 GPa for Ta impacting Pt at 7.8 km/s (Holmes et al., 1989). This range of pressure enables investigations of equations of state (EOS), dynamic strength, solid-solid phase transitions and crossovers, melting, electrical conductivity and so forth.

The piston in Fig. 3.1 has a mass as large as ~5 kg, which is large compared to 60 g of the H_2 gas plus ~25 g of a typical impactor, of which ~18 g is usually an Al, Cu, Ta or Pt plate hot-pressed into a Lexan polycarbonate sabot. Hot gases from burned gunpowder drive the 90 mm diameter ~5 kg piston down the 10 m long pump tube, compressing the H_2 gas. At a gas pressure of ~0.1 GPa, the compressed

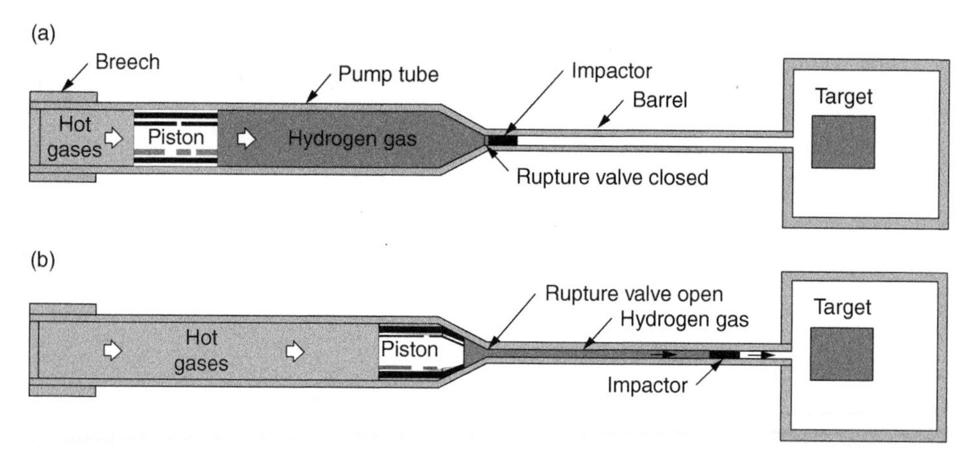

Fig. 3.1. Two-stage light-gas gun (2SG) at LLNL. First stage is hot gases from burned gunpowder plus piston; second stage is compressed H_2 gas plus impactor. Piston compresses H_2 gas in pump tube. When H_2 pressure reaches ~0.1 GPa, the rupture valve opens and isentropically compressed H_2 gas flows at ~constant rate through tapered section, which amplifies H_2 velocity as inner diameter of tapered section decreases, which accelerates impactor along 9 m barrel. Maximum impactor velocity is 8 km/s. Impact generates extreme conditions in target, which are recorded with electronic, optical and X-ray diagnostics (Nellis et al., 1981a and 1999). Copyright 1999 by American Physical Society.

H_2 gas breaks the rupture valve and flows into the narrower, evacuated barrel, 28 mm in diameter and 9 m long. The mass flow rate of the gas through the taper is ~constant because the weight of the piston is large compared to the weight of H_2 gas plus impactor. Reduction in diameter in the taper from pump tube to barrel causes gas velocity to increase to as much as 8 km/s.

Highest impactor velocity is achieved with the driving gas having the highest sound velocity c for the gas to flow as fast as possible through the taper to transmit gas pressure to the impactor at the fastest possible rate. Since the driving gas behaves essentially as an ideal gas, sound speed $c = (\gamma k_B T/M)^{0.5}$, where γ is the ratio of specific heat at constant pressure to that at constant volume, k_B is Boltzmann's constant, T is temperature of the compressed gas and M is its molecular weight. Since H_2 has the smallest M of any molecular gas, it has the highest sound velocity and, thus, achieves the highest impactor velocity and impact-shock pressure relative to any other potential driving gas. Velocity of the impactor in free-flight is determined by measuring the time between two fast X-ray pulses a measured distance apart (Mitchell and Nellis, 1981a). Photographs of the LLNL two-stage gun facility are published (Nellis, 2007a).

General Motors, Delco Electronics Division provided the 19 m long 2SG to LLNL in 1972. General Motors had used that 2SG as a test bed for their substantially larger 2SG, which was used to test thermal shielding for re-entry vehicles returning from exo-atmospheric space and for Hugoniot experiments on liquid D_2 (van Thiel et al., 1974). The larger gun enables use of substantially higher mass projectiles but little increase in velocity relative to that of the smaller 2SG. After the U.S. National Aeronautics and Space Administration (NASA) reached the moon in 1969, Delco provided one of those 19 m long guns to LLNL for scientific materials investigations at a nominal cost.

3.2 Mass Acceleration by Pulsed Power: Z Accelerator

Impact velocities as large as ~45 km/s have been achieved with the Z Accelerator at SNLA driven by magnetic pressure generated by fast, magnetic-flux compression (Hall et al., 2001; Schwarzschild, 2003; Lemke et al., 2011; Knudson and Desjarlais, 2013; Davis et al., 2014). Magnetic pressure $P(B) \propto B^2$, accelerates a metal impactor to ultrahigh velocities with magnetic field B generated by an enormous electrical current pulse I(t). Maximum current produced by Z is ~25 MA, which generates a magnetic field of ~10 MG and a magnetic pressure of several 100 GPa applied to an impactor. Impactors contain a metallic layer so that magnetic flux diffusion times through them are long compared to acceleration time of the impactor. In this way high magnetic drive pressures are maintained and eddy current heating of the impact region is minimized.

Electrical current of Z is generated with a Marx bank and several associated pairs of gas switches charged to more than 20 MJ. This source produces a current pulse with a rise time of 100 to 500 ns, which is dumped into an electrical load at the center of the Z Accelerator, which is 34 m in diameter and 7 m high. The resulting B field accelerates two metal plates in opposite directions over 3–4 mm vacuum gaps. Useful thicknesses of the Al and quartz drive plates and samples are typically a few 100 microns. A photograph of the facility on firing shows sparking, which is caused by power leakage from fast switches submerged in water (Schwarzschild, 2003).

The electrical load of Z that generates ultrahigh velocities is illustrated in Fig. 3.2. The time-dependent current-density pulse $J(t)$ from Z flows upward in the layer on the far left, which generates an azimuthal magnetic field B into the plane of the figure. Magnetic pressure $P(B)$, the cross product of J and B, is in the direction to the right in Fig. 3.2. Thermodynamic conditions induced in a material are tuned by tuning the shape of J(t). A flyer for Hugoniot experiments is made of a high-conductivity metal, which slows substantially magnetic-flux diffusion into the flyer, which is also the anode of Z. The large J x B force accelerates a flyer to very high velocity. After a flight distance of ~3 mm, the flyer impacts the target. Velocity history of the flyer and the dynamic wave induced in the target are measured with velocity interferometers (VISAR).

Veloce is a system similar to the Z Accelerator but much smaller in scale. Because of its smaller operating voltage, peak pressures and stresses are

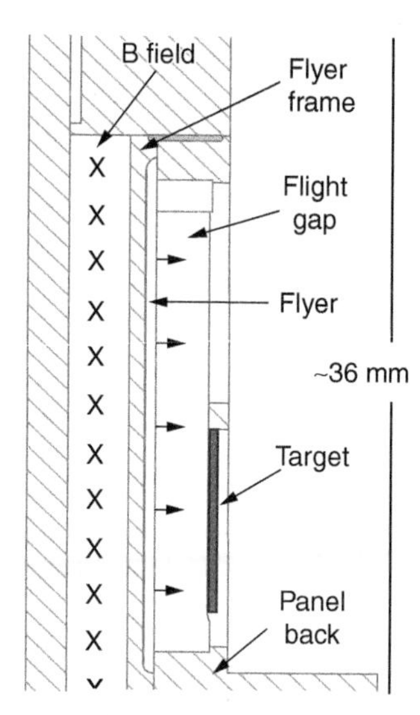

Fig. 3.2. Schematic of flyer acceleration by magnetic pressure generated by time-dependent current density J from Z Accelerator upward in layer on far left, which generates B field into plane of figure. Flyer frame is anode of Z with which flyer (impactor) is in contact and thus accelerated to right. Width of flight gap is ~3–4 mm (Knudson et al., 2003b). Copyright 2003 by American Institute of Physics.

correspondingly reduced relative to Z. Veloce is useful for isentropic and shock compression experiments on research-size scales: dynamic strength, pressure-shear measurements and a variety of other materials investigations (Ao et al., 2008; Alexander et al., 2010).

3.3 Giant Pulsed Lasers

Dynamic compression is generated by high-intensity pulsed laser irradiation at the NIF, a giant pulsed laser at Lawrence Livermore National Laboratory. Giant pulsed lasers also exist at the Omega laser at the University of Rochester and at the Linac Coherent Light Source at SLAC National Accelerator Laboratory. Many shock experiments are performed with few ns laser pulses with diagnosis of the resulting decaying shock wave (Fig. 2.11). A wide variety of physical properties have been measured (Bradley et al., 2004; Eggert et al., 2008, 2009; Smith et al., 2014; Gorman et al., 2015). Ramp-compression research includes measuring quasi-isentropes up to pressures of several 100 GPa for comparison with 0-K isotherms calculated theoretically.

At the University of Illinois, a small-scale laser-driven mass accelerator has been developed with diagnostics that combine fast optical spectroscopy/microscopy with photon Doppler velocimetry. A laser accelerates 0.5 mm diameter Al or Cu flyer plates to velocities as high as 6 km/s. High shock pressure is generated on impact with a glass slide. Fast optical spectroscopies and chemical reactions are investigated (Banishev et al., 2016).

3.4 Quasi-Isentropic Cylindrical and Spherical Compressions

Since the 1960s quasi-isentropic and shock compression experiments on hydrogen isotopes have been performed with cylindrical and spherical implosion systems driven by high explosives (HE). Radial convergence is utilized in both single- and two-stage implosion systems. In a two-stage implosion a seed magnetic field is injected from an external source. Convergence compresses magnetic flux, which increases magnetic drive pressure. That is, two-stage implosions use HE to dynamically compress an interior metallic shell, which in turn compresses interior magnetic flux against a second innermost metallic shell. Magnetic pressure generated by increasing magnetic field in the first stage then compresses an interior sample of a gaseous, liquid or solid hydrogen isotopes in the second stage (Zhernokhletov et al., 1995).

The purpose of the magnetic-flux compression stage is to essentially isolate the hydrogen sample at the center from shock dissipation generated by HE in the outermost region. With the two-stage system, magnetic pressure is gradually applied to the innermost metallic shell, which means pressure is also applied gradually to the sample in the central region. In this case, shock dissipation is virtually eliminated from the compression, which is then virtually isentropic.

In the single-stage implosion, HE is in contact with a single metallic shell, which means the sample in the central region experiences an initial shock wave from the HE, which shock-heats the sample to some extent prior to subsequent quasi-isentropic compression as the shell coasts radially inward after the initial shock.

These experiments are based on the assumptions that (1) sample density can be determined with flash X-radiography to measure radial positions of dense solid shells containing a hydrogen isotope and (2) pressure history can be calculated with hydrodynamic simulations of the process. Hawke et al. (1978) performed pioneering experiments on isentropic compression of hydrogen and neon. Trunin et al. (2010) have published an extensive review of a variety of experiments on hydrogen isotopes.

3.5 Static Compression: Diamond Anvil Cell

Static compression produces states at high pressures with lifetimes that are long compared to time required to apply pressure and long for heat produced by static compression to diffuse out of a sample at the speed of sound. For this reason static compression is slow, isothermal and generates relatively little entropy, although static-compression-induced disorder is commonly observed at high static pressures by broadening of X-ray diffraction and optical spectroscopic lines measured in diamond-anvil cells (DAC)s. Hugoniot curves are used to derive pressure calibrations, called shock-wave reduced isotherms, to determine pressure up to ~200 GPa at 300 K achieved in static compression experiments (Chijioke et al., 2005b).

3.5.1 Diamond anvil cell (DAC)

Highest static pressures are achieved with a diamond anvil cell (DAC), illustrated in Fig. 3.3. An extensive body of research is conducted under static pressures extending from ~10 GPa to pressures up to ~600 GPa (Piermarini and Block, 1958; Jayaraman, 1983, 1984; Ruoff et al., 1990; Bassett, 2009; Dubrovinsky et al., 2012). The intrinsic difference between static and dynamic compression is the rate at which pressure is applied. Static pressure is typically applied in ~s and experimental lifetimes usually range from seconds to months, depending on sample material, *P*, *T*, chemical diffusion, chemical corrosion/reactions, etc.

Sample heating in a DAC is typically done by laser-heating, by passing an electrical current through a conductor in the sample volume or by heating the entire

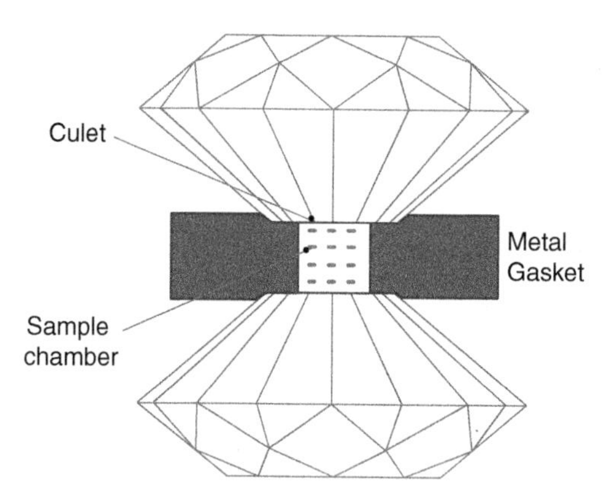

Fig. 3.3. Diamond anvil cell (DAC). Sample is contained in cylindrical chamber at center of metal gasket, which is compressed between two gem-quality diamonds to pressures as high as 300–500 GPa. To achieve highest pressures sample is initially ~20 μm in diameter and ~5 μm high or so. Mechanical assembly to constrain and maintain alignment on compression is not shown. Because diamond is optically transparent, lasers with tiny spot sizes are common spectroscopic probes, as well as X-ray beams. Insertion of electrical leads to measure electrical conductivity is common at lower pressures (Jayaraman, 1983, 1984). Copyright Reviews of Modern Physics 1983.

DAC. Sample heating and cooling in a DAC is independent of the compression process, unlike dynamic compression in which heating and entropy are produced as part of the fast, adiabatic dynamic compression process itself. Static compression techniques and experimental results have been reviewed (Mao and Hemley, 1994; Eremets, 1996; Hemley et al., 2002).

3.5.2 Static Pressure Calibration

A key problem is the determination of pressure achieved in a DAC. As of yet there is no known primary static-pressure standard above ~5 GPa (Bean et al., 1982, 1986). Calibration of pressure achieved under static compression has been a major issue ever since Bancroft et al. (1956) reported a phase transition in Fe at a shock pressure of 13 GPa and estimated temperature of 40°C. In 1961 a phase transition was detected in Fe at around 13 GPa static pressures by electrical resistance measurements (Drickamer and Balchan, 1961). In 1962 a phase transition was observed in Fe by X-ray diffraction up to ~15 GPa (Jamieson and Lawson, 1962). The phase above 13 GPa was deduced to be ε-hcp, which is described by two lattice parameters. The hcp phase determination was based on observation of only one xrd line, which was *atypical* of the initial bcc Fe α phase, plus the constraint that the higher-pressure phase above 13 GPa under static compression should have the volume of the high-pressure phase measured by Bancroft et al. under shock compression. Thus, Hugoniot measurements determined the pressure and density of that Fe ε-hcp phase transition and the single measured xrd line together with measured shock density constrained the observed xrd line to be indexed to that of the ε-hcp phase. The pressure of the α−ε Fe phase transition is discussed in more detail in Chapter 4.

There are many secondary static-pressure standards. The commonly used Ruby scale is calibrated versus pressures derived from shock-wave reduced isotherms (Mao et al., 1978; Wang et al., 2002; Chijioke et al., 2005a, 2005b). In addition, isotherms calculated theoretically are often assumed to be correct absolutely and used as pressure standards (Kunc et al., 2003; Holzapfel, 2005, 2010; Dorogokupets et al., 2012). However, calculated "theoretical" pressure standards have intrinsic systematic uncertainties and provide reasonable estimates of pressure.

Shock-wave reduced isotherms are derived from measured Hugoniots obtained with absolutely accurate Eqs. (2.1) to (2.3). Hugoniots are then corrected to pressures and volumes at, say 300 K, by assuming thermal pressures calculated with a density-dependent Gruneisen model and estimating shock-induced stress contributions to shock pressures caused by strength (Chijioke et al., 2005a, 2005b). Use of a Gruneisen model and strength contributions introduce systematic uncertainty into the calculations of a pressure standard. To minimize effects of these corrections, the sum of such corrections of a given material is often restricted to values that are small compared to shock pressure at a given density, say 30% of P_H at given V_H.

Correcting Hugoniot data, which are functions of density and temperature, to obtain a 300 K isotherm becomes increasingly more uncertain as shock pressure increases above ~200 GPa, depending on material. Shock temperatures increase more rapidly with shock pressure for lower Z materials, such as Al, than for higher Z materials, such as W. As shock pressure increases, the Gruneisen parameter becomes a function of temperature as well as density, though the relationship is not known. As shock pressure increases strength contributions to Hugoniot pressure decrease, though the relationship is not known. As shock pressure increases, shock compression approaches a limiting value of approximately fourfold but the 300 K isotherm has no limiting compression. For all these reasons, the use of the Gruneisen model to obtain 300 K isotherms from Hugoniots becomes systematically less reliable with increasing pressure. Because more accurate calibrations might become available in the future, it is important to state exactly how static pressures are determined in given static-pressure experiments so that corrections can be made in the future, as appropriate.

A likely direction in which to proceed to develop isothermal static-pressure standards above ~200 GPa is to correct measured dynamic quasi-isentropes or ramp waves (Nellis, 2007b; Smith et al., 2014) to obtain static isotherms to pressures above ~600 GPa as estimated, which is achieved in a two-stage diamond-anvil cell (Dubrovinsky et al., 2012). Such corrections to dynamic quasi-isentropes or ramp waves are analogous to corrections to Hugoniots to achieve static-pressure isotherms up to ~200 GPa. However, temperatures achieved by ramp-wave and QI compressions are substantially lower than achieved by shock compression. In contrast, stresses induced by strength under QI compression above 200 GPa might be substantial and larger than those on the Hugoniot. Determination of static pressures in the 500 GPa regime is an important issue because static pressures in this regime are starting to be accessed experimentally.

4

Brief History of High-Pressure Research: 1643 to 1968

Over the past four centuries high-pressure research has amassed a distinguished record of accomplishment ranging from (1) measurement of the pressure that Earth's atmosphere exerts on Earth's surface, which proved the atmosphere has mass, a major question in 1640, to (2) state-of-the-art experiments and theory on the virtually universal nature of atomic matter at extremely high pressures, densities and temperatures. Today the issue of whether Earth's atmosphere has mass or not seems difficult to accept as ever having been a serious question. However, in the seventeenth century no less a scientist than Galileo thought Earth's atmosphere to be weightless, which motivated invention of the barometer and the first measurement of the "weight" of the atmosphere by a scientifically curious student named Torricelli, who did not believe the opinion of his professor, Galileo.

An experimental science begins when a property can be measured with sufficient accuracy to draw a significant conclusion about a scientific question. The key requisite is an experimental measurement and its estimated accuracy. Modern high-pressure research developed into its present form with dynamic and static compression from 1643 to 1961. The criteria for the choices of these dates is the first quantitative measurement of pressure in 1643 and general acceptance by the static high-pressure community in ~1961 that dynamic high-pressure research is a science based on experimental detection under both static and dynamic compression of the $\alpha-\varepsilon$ transition in Fe at a pressure of 13 GPa.

Up until ~1850 only static pressures had been investigated. In the last half of the nineteenth century the basic idea of a shock wave was developed theoretically by W. J. M. Rankine of the University of Glasgow (1870). Rankine's theoretical work on shock compression began with completion of the derivation of the ideal-gas EOS, $PV = RT = (2/3)E$, where R is the gas constant. In 1848 Lord Kelvin, also of the University of Glasgow, devised a temperature scale with an absolute zero, which is independent of the properties of any particular substance and is based on Carnot's theory of heat. By international agreement, absolute 0 K is taken as -273.15° C.

Rankine has been called a founding contributor to thermodynamics, along with William Thompson (Lord Kelvin) and Rudolph Clausius.

Dynamic high-pressure experiments began in earnest in the 1940s as a result of World War II. Not until 1961 was it generally accepted by the static high-pressure community that dynamic compression is a science – that is, that experimental results of dynamic compression were being interpreted correctly. The main concern was the apparently brief timescale on which shock experiments occur (< μsec) compared to "human" timescales of ~sec.

In 1956 researchers at Los Alamos published the first shock-induced phase transition in Fe, now known as the Fe $\alpha-\varepsilon$ transition at 13 GPa (Bancroft et al., 1956). Shock pressures were generated with high explosives. When P. W. Bridgman, a Nobel Laureate, heard this result, he measured the electrical resistance of Fe versus static pressure. When he found no indication of a transition near what he thought was 13 GPa in his experiments, Bridgman said it was unlikely that a crystal could change phase in a ~μsec (Bridgman, 1956). To test Bridgman's assertion H. G. Drickhamer and A. S. Balchan (1961) and J. C. Jamieson and A. W. Lawson (1962) both spent ~five years building apparatus to measure electrical resistance and X-ray diffraction, respectively, above 10 GPa in a large-volume press to look for a resistance anomaly and lattice diffraction, respectively, at static pressures up to ~15 GPa. The fact that Fe actually undergoes the $\alpha-\varepsilon$ transition in a time as brief as a ns under dynamic compression was eventually observed in 2005 (Kalantar et al., 2005).

4.1 Evangelista Torricelli: 1643

Torricelli made the first quantitative measurement of pressure in 1643. To do so, he invented the Hg barometer to see if Earth's atmosphere has weight – that is, if the atmosphere exerts pressure vertically downward on Earth's surface. Torricelli was a student of Galileo, and Galileo claimed the atmosphere does not have weight. Torricelli thought it did based on some then puzzling engineering observations, for which he had a proposed explanation, and set out to test his idea. Torricelli eventually invented the Hg barometer, measured the value of atmospheric pressure and reportedly found atmospheric pressure varies with elevation by running up and down hills in central Italy. In so doing so, Torricelli experienced the thrill of discovery and founded high-pressure science.

The existence of forces exerted by air had been well known to sailors and homeowners for millennia. However, Galileo and most of the few other scientists of those days thought that air exerts forces only in directions parallel to the Earth's surface and thus is massless – ethereal one might say. This was not an unreasonable assumption, given that the human body is unable to detect atmospheric pressure. Effects of wind were then essentially the only observations that indicated Earth has an atmosphere. At that time there was no existing theory applicable to the question – only opinion. The question of whether or not air has mass, and thus weight, needed an experiment to settle the issue.

The basis of Torricelli's opinion that air has weight was the fact that engineers could use a water siphon to lift materials to a limiting height of ~34 feet and no more. Torricelli thought that atmospheric pressure might explain the height limit of a water siphon. He knew that the density of liquid Hg is about 14 times greater than that of water and so reasoned that, if total weight/area of a liquid column is all that matters, then the atmosphere should support a column of Hg about ~30 inches high. So Torricelli built a glass barometer and observed that when a pool of liquid Hg was left open to the atmosphere and an evacuated long glass tube, sealed on one end and open on the other, was inserted into the Hg pool with its axis vertical to the Hg surface, liquid Hg rose in the evacuated tube to a height of ~30 inches. His barometer verified his prediction. After observing that his intuition was correct, Torricelli is reported to have said enthusiastically, "We live submerged at the bottom of an ocean of air." He is also said to have walked up and down hills and observed ~mm changes in the height of his Hg column caused by changes in atmospheric pressure with elevation. Torricelli had experienced the thrill of discovery.

Torricelli's barometer is so sensitive, accurate, simple and useful that it is still used today to measure barometric pressure. Named in his memory, a torr is the unit of pressure required to raise the level of a Hg column 1 mm. The torr is commonly used to quantify pressures from $\sim 10^{-12}$ torr up to $\sim 10^{3}$ torr.

Torricelli went into a region that was a scientific frontier of his day. He went where no experimental scientist had gone previously, developed a new diagnostic with great sensitivity and resolution to look for something he thought might be there and, with his new instrument, resolved a major scientific controversy. Nature does not always behave in the way one might expect, not even for a great scientist like Galileo. Experiments, not opinion, are required to prove a scientific hypothesis or theoretical prediction. This idea is known as the scientific method.

Galileo himself experienced the thrill of discovery several times. In 1610, by peering into his crude ~10x magnification telescope, he observed faint images of the Galilean satellites, the four largest moons of Jupiter. One can only imagine the thrill Galileo must have felt in realizing that he was the first person on Earth to learn that other planets have moons. Materials in the deep interiors of the Galilean satellites, in Jupiter, in Saturn and in other planets within and beyond our solar system are themselves under very high gravitational pressures and are studied today at planetary pressures and temperatures obtained in laboratories.

With his questioning mind and discoveries, Galileo played an important, if indirect, role in the establishment of high-pressure research and of science in general.

4.2 Blaise Pascal: Experimental Verification

Later in the 1640s Blaise Pascal also constructed a barometer. Pascal measured barometric pressure as a function of altitude and verified Torricelli's observations of barometric pressure. Pascal also got involved in a controversy of those days as to whether or not a vacuum can exist, as for example, above the Hg column in his barometer. Today, the MKS unit of pressure is the Pascal (Pa). MKS units are based on length in meters, mass in kilograms and time in seconds.

4.3 Ideal-Gas Equation of State: 1660 to 1848

Development of the ideal-gas EOS was the next important step in the development of dynamic high-pressure research. An EOS is a relation between thermodynamic variables, such as pressure, density and temperature. Rankine used the ideal-gas EOS explicitly for the first derivation of the equations of conservation of momentum, mass and internal energy across the front of a shock wave. The ideal gas EOS is used today to demonstrate basic ideas of dynamic compression, such as limiting shock compression and how many weak shocks are required for a quasi-isentrope to be an isentrope from the first-shock state of a multiple-shock compression (Section 2.1.7).

The EOS of an ideal gas is PV = RT. It took almost 200 years after P was first measured to derive this simple form of the EOS for several reasons, not the least of which was a general belief that the expected simplicity of nature required that P, V and T must each be measured relative to their respective absolute zeroes. An ideal gas is one that satisfies several assumptions: (1) particles are point masses with zero volume per particle; (2) interactions between particles are negligible, (3) pressure is caused only by elastic collisions of particles with walls containing the gas and (4) temperature is caused by kinetic energy of particles in the gas. $P = 0$ in a vacuum because there are no particles to impart momentum to walls. $V = 0$ in a vacuum because there are no particles at all.

But back in the seventeenth to nineteenth centuries, the meaning of $T = 0$ was unknown. It was realized at that time that absolute zero T might exist but experiments had yet to be performed at very low temperatures to give clues as to the nature of matter there. Temperatures near absolute zero were not actually achieved until 1898 and 1908 (20 K in liquid H_2 and 4.2 K in liquid He) by James Dewar and Kamerlingh Onnes, respectively (Mendelssohn, 1966). Today we know there is an absolute zero of temperature and that Lord Kelvin deduced absolute zero temperature in degrees C in the middle of the nineteenth century (Thomson, 1848). However, in the seventeenth century, whether absolute zero temperature existed had yet to be determined and an absolute temperature scale had yet to be developed.

Based on history of that period, one can speculate on what happened after Torricelli invented a technique to measure gas pressure. After 1643 it was possible to determine experimentally the relation between absolute P and V of a gas at a fixed temperature over a limited range. In those days the height of a Hg column in a barometer and linear dimensions of a sealed box containing air could probably be measured to a ~mm or better with a "ruler". Pressure changes of an atmosphere or so could probably be effected with then available pumps and seals. Thus, the relation between pressure and volume could be measured with sufficient accuracy to demonstrate a quantitative relation between pressure and volume of air, an ideal gas, at room temperature. In 1662, nineteen years after the invention of the barometer, Boyle reported the well-known law named after him, that pressure and volume of an ideal gas are inversely proportional to one another at fixed temperature. Boyle's law applies to all ideal gases at fixed temperature, not simply to air.

One might then ask the question, why did it take so long to go from the invention of the barometer in Italy to the discovery of Boyle's law in England? Given the slow rate of diffusion of scientific information from Italy to England and elsewhere in those days, the small number of scientists, the time required for Boyle to think about an interesting experiment to do with the new pressure-measuring instrument, the time it took Robert Hooke, Boyle's assistant, to build Boyle's apparatus, the political turmoil in England that followed when King Charles I was deposed in 1642 by Oliver Cromwell, the execution of Charles I in 1649 and the restoration of Charles II to the monarchy in 1660, nineteen years is a relatively short time for Boyle to come up with his law.

The next advance in the science of ideal gases, the temperature dependence of $P(V)$, took much more than two decades to discover. Charles' law, in its present form, states that the volume of an ideal gas is inversely proportional to its absolute temperature at fixed pressure. In 1787 Jacques Charles found experimentally that in a given temperature interval of about 80 K, air, hydrogen, oxygen, nitrogen and carbon dioxide expand essentially the same amount over the same temperature interval. Charles recorded his observations in his notebook but never published his results. In 1802 Joseph Guy-Lussac obtained similar experimental results, which indicated a linear relationship between volume and temperature, and he credited the discovery to the previous unpublished work of Charles.

In 1834 Emile Clapeyron combined Boyle's and Charles' laws into what today is called the Ideal Gas law. Temperature was in units of C. Thus, to express temperature relative to absolute zero, it was essential to add a constant temperature, the magnitude of absolute zero, which in 1834 was thought from experiments to be −267° C. In 1848 William Thomson, Lord Kelvin, of the University of Glasgow, derived an absolute temperature scale on which the value of absolute 0 K on the centigrade scale was derived from the laws of thermodynamics to be 0 K = −270° C (Thomson, 1848), near the value derived experimentally in those days using the absolute temperature scale of the "air" thermometer and near the now internationally accepted value of 0 K = −273.15° C. Thus, in 1848 the derivation of the Ideal Gas law was completed, 200 years after the first measurement of pressure by Torricelli.

4.4 Theoretical Concept of a Shock Wave: 1848 to 1910

The starting point of the conceptual development of shock flows was the Navier-Stokes equation, named after C.-L. Navier and G. G. Stokes (Liu, 1986). Claude-Louis Navier was a French engineer and physicist whose primary interest was mechanics. He was Inspector General of the Corps of Bridges and Roads. Navier was admitted into the French Academy of Science in 1824. In 1830 he became professor at the École Nationale des Ponts et Chaussées. In 1831 he succeeded A. L. Cauchy as professor of Calculus and Mechanics at the École Polytechnique. George Gabriel Stokes was a mathematician and physicist at Cambridge University, who made important contributions to fluid dynamics, optics and mathematical physics. Stokes was a Fellow, Secretary, and President of the Royal Society of London. Major contributors to the development of the idea of a shock wave were J. Challis, E. E. Stokes, S. Earnshaw, B. Riemann, W. J. M. Rankine, P.-H. Hugoniot and J. W. Strutt (Lord Rayleigh) (Courant and Friedrichs, 1948). The Ideal Gas law was completed by the time the concept of a shock wave began to be developed and played an important role in the development of shock hydrodynamics.

Challis tried to solve a differential equation for flow of an isothermal gas in terms of a simple wave but found that it was not always possible to find a solution with a unique wave velocity (Challis, 1848). Stokes proposed that a discontinuity in wave velocity occurs when the rate of change of wave velocity with run distance becomes infinite (Stokes, 1848). He also argued that discontinuities in wave velocity cannot occur in real systems because such a discontinuity would be smoothed by viscous forces. Earnshaw developed a simple solution to a wave equation for matter in which pressure is a function only of density, $P = P(\rho)$. For a compression wave with $dP/d\rho > 0$, sound velocity increases with P. Thus, an initial tendency of pressure to increase with density would grow larger with density until the front of a sound wave would steepen into a discontinuity called a shock wave (Earnshaw, 1860). Riemann expanded on previous solutions of the flow problem, incorrectly assuming that the transition across a shock front is both adiabatic and reversible (Riemann, 1860).

W. J. M. Rankine was a Scottish physicist, engineer, and Fellow of the Royal Society of London. In his early years he was one of the major contributors to the new field of thermodynamics, along with Clausius and Lord Kelvin. From 1855 to 1872 Rankine was a professor of civil engineering and mechanics at the University of Glasgow. In 1859 W. J. M. Rankine proposed the absolute Rankine scale of temperature in which temperature is expressed in R, rather than in K, relative to absolute zero. A degree R is identical to a degree F. The R scale is used in engineering applications. Rankine did substantial research in thermodynamics of gases.

Rankine and later Hugoniot developed conservation equations for momentum, mass and energy across a shock front, which compression is adiabatic with respect to heat transfer from outside the front while thermal energy may be exchanged by particles within the shock front (Rankine, 1870). That is to say, thermal equilibrium can be achieved under adiabatic compression within a shock front. Rankine derived his conservation equations specifically using the EOS of an ideal gas.

P.-H. Hugoniot was a member of the French marine artillery service in which he was professor of mechanics and ballistics, then Assistant Director of the Central Laboratory of Marine Artillery and finally Captain of Marine Artillery. His research received an award of the Paris Academy of Sciences in 1884 (Cheret, 1992). Hugoniot, like Rankine, derived conservation equations of momentum, mass and energy across a shock front. Those equations are now called the Rankine-Hugoniot (RH) equations.

Hugoniot showed that conservation of energy implies an entropy change across a shock front. Hugoniot (1887, 1889) and Strutt (Lord Rayleigh) (1910) both pointed out that a shock front cannot be both adiabatic and reversible, as Riemann had claimed, because of conservation of energy. Since shock compression is adiabatic, it is irreversible. Rayleigh also pointed out that because entropy must increase across a shock front, isentropic release of shock pressure by a rarefaction shock (a sharp discontinuous drop in pressure) cannot occur in an ideal gas, which is a single-phase material. Rarefaction shocks are observed to occur at phase transitions (Grady, 1998).

Lord Rayleigh had a broad interest in hydrodynamic instabilities in various types of fluid flows. One such flow was the instability developed in layers of fluids of variable density (Strutt, 1883). In particular, he considered instability growth of an interface between two fluids of different densities in which the light fluid pushes on the heavy fluid, which results in turbulent mixing of the two materials along that interface. Shock-induced mixing at the interface between a fuel capsule and an enclosed D-T fuel ball is an important limiting issue today in ICF.

Mathematical development of the concept of a shock wave was essentially completed in 1910 (Strutt, 1910). That seminal theoretical work was stored on library shelves in Europe for several decades awaiting experimental verification, which was not possible in those early days. A major motivation was needed to fund experimental development of the investigation of shock propagation.

4.5 In the Beginning: Early 1940s

Shock flows are supersonic and very fast. Fast cameras, fast electronic diagnostics, short-pulse X-radiography, fast triggers and facilities to generate high shock pressures with explosive shock drivers are required for generating and diagnosing shock experiments in condensed matter. In 1900 those fast diagnostics did not exist and their development was well beyond technological capabilities of those days. A major motivation was needed for governments to provide the substantial funding, facilities and technical staff required for such a shock driver and diagnostic development program.

Whereas the Ideal Gas law was derived virtually entirely from experimental data, the idea of a shock wave was developed in three distinct phases. In the first phase during the last half of the nineteenth century, the concept of a shock wave was developed in Europe from the theory of hydrodynamics and the interrelation between supersonic fluid dynamics (nonlinear mechanical flows) and thermodynamics (Courant and Friedrichs, 1948). That is, dynamic pressures, densities, internal energies and entropies are achieved by supersonic nonlinear flows, and the nature of such a flow depends on the response of the particular material that is flowing via its EOS (Bethe, 1942).

The second major phase of the development of shock-wave research began near the beginning of World War II. At that time, development of a picture of the equilibration process in the front of a shock wave was initiated (Bethe and Teller, 1940). This work enabled Bethe and Teller to become part of the Manhattan Project, whose purpose was to build the first atomic bomb. Bethe and Teller had immigrated to the United States from Germany in the 1930s and very much wanted to be part of America's war effort. However, in 1940 they were still classified as enemy aliens (non-citizens), who where forbidden by Congress to work on the Manhattan Project. After they developed their 1940 calculation on the nature of thermal equilibration in the front of a shock wave, their unpublished calculation was declared classified material, and Bethe and Teller were permitted to join the Manhattan Project at Los Alamos as group leaders.

World War II generated substantial governmental funding for experimental facilities to test theoretical predictions about shock-wave propagation. A major emphasis of that period was the development of fast experimental techniques to diagnose shock flows, particularly measurements of shock-compression *P-V* data using the R-H conservation equations.

The third phase of the development of shock-wave research, from the early 1950s to 1961, emphasized understanding the relation between experimental results obtained with shock and with static compression. This issue was brought on by the fact that both shock and static research were going into regimes of higher pressures than investigated previously. During this period, pressure standards were a major issue for static-pressure research. In contrast P_H, V_H and E_H are obtained experimentally in dynamic-compression experiments by the R-H conservation equations across the front of a shock wave.

4.6 Experimental Development of Supersonic Hydrodynamics: 1940s to 1956

With the arrival of World War II, interest in shock compression increased exponentially and rapidly. While Rankine and Hugoniot had both said energy is exchanged between particles within a shock front and, thus, thermal equilibration is possible, the fact that thermal equilibrium is reached in a shock front had yet to be demonstrated. In gases, liquids and most solids, the thickness of a shock front is the thickness of the region at the front of a shock wave in which material comes into thermal equilibrium. Bethe and Teller showed that values of thermodynamic parameters behind a shock front are uniquely determined by their values ahead of the shock front, independent of the path taken toward thermal equilibrium in the extremely non-equilibrium conditions in the width of a shock front (Bethe and Teller, 1940). Hans Bethe went on to win the Nobel Prize in Physics in 1967 for his research in nuclear physics. That 1940 research was never published in a scientific journal because it was written during wartime. However, Bethe considered that paper to be in the top 10% in his "Selected Works" list. Today that paper is a key work in the study of solids far from equilibrium (Mermin and Ashcroft, 2006).

The motivation to develop shock wave experimental facilities was provided by the onset of World War II. Shock compression experiments in the United States began in the Manhattan Project, whose goal was the development of nuclear weapons. The Manhattan Project was built on the future site of Los Alamos National Laboratory. A similar project was undertaken in the Soviet Union.

In 1944 electrical detectors, called "pins", were in use at Los Alamos to measure shock wave velocities in condensed matter. Electrical pins measure arrival times at discrete points in space at the front of a travelling shock wave. By suitably averaging arrival times at known positions, shock velocity u_s can be determined. Shock velocity u_s is then used to calculate mass velocity u_p behind the shock front with shock-impedance matching. Knowledge of u_s and u_p enables the calculation of shock pressure, specific volume, and specific internal energy, P_H, V_H, and E_H, respectively, via the R-H equations.

By 1950 the optical flash-gap method (McQueen et al., 1970) was developed to measure shock arrival times for the same purpose as electrical pins. In the flash-gap method, a thin layer of Ar gas is placed between two thin solid shims, one of which is transparent. When a strong shock wave transits the Ar gas, the gas is shock-compressed sufficiently to produce hot plasma and a sharp, brief, optical flash. A rotating-mirror streak camera views an assembly and records shock arrival times at the various flash gaps arrayed in space.

In the mid-1950s a large number of experimental results obtained in the late 1940s and early 1950s started to appear in the scientific literature. Since then American and Russian defense laboratories have published an enormous amount of experimental shock results. Subsequent to World War II, France, Germany, Japan, China and several universities have also established shock-compression experimental research programs. With these experimental facilities, theoretical development of shock-wave physics from ~1850 to 1942 were demonstrated to be correct by experiments performed in the last half of the twentieth century. That theoretical development, essentially without benefit of experimental data, was a major theoretical accomplishment.

In 1950 G. I. Taylor published his theory on instabilities of interfaces between different materials under steady acceleration (Taylor, 1950). His work was complementary to that of Lord Rayleigh (Strutt, 1883). Today such interfacial instabilities are known as Rayleigh-Taylor instabilities and are investigated widely in a large number of fluid-mechanics problems.

4.7 P. W. Bridgman's Contributions to Dynamic Compression: 1956 to 1961

P. W. Bridgman founded modern static high-pressure research (Nellis, 2010). Static high-pressure experimental research began in European universities in the last part of the nineteenth century. Bridgman conducted extensive static high-pressure experiments for more than fifty years at Harvard University. His research is well documented in seven volumes of his collected works (Bridgman, 1964).

In the last five years of his career, Bridgman made significant contributions to the scientific foundations of dynamic compression as well. His activities included statements about shock compression of Fe that motivated static researchers to build systems to perform X-ray diffraction experiments and to measure electrical resistances of Fe at pressures up to ~15 GPa. The goals of those static-pressure experiments were (1) to demonstrate that the $\alpha-\varepsilon$ transition in Fe does occur under static compression as well as shock compression at 13 GPa and (2) to determine the likely crystal structure of ε-Fe. Bridgman also made a prediction that motivated Soviet researchers at Arzamas-16 to perform the first experiments at ultrahigh shock pressures in proximity to underground nuclear explosions.

Bridgman was Professor of Physics at Harvard. In 1946 he won the Nobel Prize in Physics for his static high-pressure experiments. His graduate students include John Van Vleck (graduated 1923), Francis Birch (1932) and Gerald Holton (1948). All three were professors at Harvard. Van Vleck was awarded the 1977 Nobel Prize in Physics for his research on magnetism in solids, a prize he shared with Philip Anderson and Sir Neville Mott. Birch was Professor of Geophysics and derived the first equation of state for materials at 100 GPa pressures in the deep Earth. Holton is Professor of Physics and the History of Science, Emeritus. Simultaneous with his high-pressure research, Bridgman had a very productive career in the philosophy of science and published several textbooks on thermodynamics and the philosophy of science (Bridgman, 1935).

Bridgman's significant involvement with shock waves began in 1956 when Bancroft et al. (1956) reported a transition in Fe to a then unknown phase at a shock pressure of 13 GPa. Shock pressures in those experiments at Los Alamos were driven by chemical explosives, and experimental lifetimes were ~ μsec. Because the calculated shock temperature of Fe at 13 GPa is only 40° C, Bridgman tried with his static pressure system to reproduce, in electrical resistance measurements, the shock-induced transition in Fe but was not able to do so. Bridgman's reaction to the achievement of Bancroft et al. was simply to state that it was unlikely for a phase transition to occur in a μsec and that whatever it was that caused that observation was not a phase transition (Bridgman, 1956).

Static-pressure researchers noted what Bridgman said and embarked on projects to measure electrical resistance (Drickhamer and Balchan, 1961) and X-ray diffraction (xrd) spectra of Fe (Jamieson and Lawson, 1962) at static pressures up to ~15 GPa. Their goal was to determine whether or not the Fe transition occurs at 13 GPa and, if it does, to determine the crystal structure of the high-pressure ε-phase. Their goal was not unlike Torricelli's with respect to whether the atmosphere has mass, namely, to test Galileo's opinion on the matter.

The capability to measure xrd patterns at 15 GPa did not exist in the early 1950s. Thus, Jamieson and Lawson designed and built a system to make those xrd measurements. After five years, they discovered that by comparing the previous shock results of Bankcroft et al. with their diffraction results that Fe does undergo a phase transition at 13 GPa. Drickhamer and Balchan (1961) also observed a phase transition in Fe at 13 GPa.

To get a bit ahead of this story, once Bridgman realized in 1961 that chemical explosives produced reliable shock-compression data, he said that nuclear explosives would be even better and predicted that highest shock pressures would eventually be achieved with nuclear explosives. L. V. Altshuler read Bridgman's 1961 prediction and in the late 1960s measured shock pressures of several TPa in proximity to underground nuclear explosions (Altshuler et al., 1968). Bridgman never learned of Altshuler's accomplishment, having passed away in 1961.

P. W. Bridgman entered Harvard University in 1900, where he studied physics through to his PhD in 1908. His first major paper was on the measurement of high pressure (Bridgman, 1909). Between 1905 and 1961, Bridgman developed experimental techniques and measured physical properties of many materials at static pressures that ranged up toward 20 GPa. During his career he published more than 200 papers.

During World War II, Bridgman performed static high-pressure experiments as part of the Manhattan Project. He discovered the $\alpha-\beta$ transition in Pu and estimated that it occurs at a pressure very approximately to be ~0.1 GPa (Bridgman, 1959). Bridgman made those Pu measurements at the Watertown Arsenal of the U.S. Army in Watertown, Massachusetts, not far from Harvard University. Today the Watertown Arsenal is but a distant memory, having been replaced by a shopping center years ago.

After World War II, Los Alamos performed shock-compression equation-of-state (Hugoniot) experiments on many materials, generating shock pressures up to several tens of GPa. Shock waves were planar and generated with chemical explosives. The detectors were either electrical pins, which measured arrival times of a shock-wave front at various spatial points in a sample (Bancroft et al., 1956), or flash gaps observed as a function of time with a rotating-mirror streak camera (McQueen et al., 1970). Experimental lifetimes were ~μsec and detector resolution was ~0.01 μsec. By analyzing measured arrival times at various spatial locations in a sample, the temporal structure of a shock wave and velocities of individual components of a multiple-shock wave could be derived.

In 1956, Bancroft and colleagues measured the Hugoniot of Fe up to ~20 GPa using ~65 electrical pins on each of several shots. Shock velocity u_s was measured and particle velocity u_p was derived from measured u_s and ρ_0, the initial Fe sample density. Shock pressure and density were calculated from u_s, u_p and ρ_0 using the R-H equations (Chapter 2). Those experiments were self-calibrating in that absolute pressure and density are given by the R-H equations.

A large number of pins were used because two and possibly three shock waves were expected in Fe, one caused by elastic strength, one possibly caused by a phase transition and one caused by plastic compression above the elastic limit at a stress called the Hugoniot elastic limit (HEL). Those three waves were expected to have different shock velocities, which means those three waves would be well separated in time in a sufficiently thick sample. Three distinct shock waves were observed, one of which was attributed to a phase transition at 13 GPa and 40° C from α-bcc Fe at ambient to a then unknown high-pressure phase. Because of the modest rise in temperature, in principal the transition reported by Bancroft et al. (1956) could be verified by measurements under static high pressures at room temperature.

Bridgman tried to verify the existence of the Fe phase transition by measuring electrical resistance of Fe up to what he calculated to be 17 GPa. He found no indication of a phase transition in electrical resistance measurements at any pressure. While acknowledging the resistance method was not definitive, he said the observation of a third shock wave probably needed to be explained by something other than a phase transition. However, it is important to realize that in the 1950s absolute static-pressure scales were quite uncertain (Decker et al., 1972; Graham, 1994). Thus, there is a very reasonable possibility that Bridgman's stated pressure of 17 GPa was substantially overestimated as suggested by Ruoff (Cornell University, private communication, 2010).

In the early 1960s two different static-pressure experiments on the same Fe sample were reported, which demonstrated the α−ε transition in Fe at 13 GPa at room temperature does in fact occur. In 1961 Balchan and Drickamer measured the electrical resistance of Fe and detected a phase transition at 13 GPa (Drickamer and Balchan, 1961), the same pressure reported by Bancroft et al. In 1962 Jamieson and Lawson reported development of the first high-pressure cell in which xrd patterns were measured in solid specimens under quasi-hydrostatic high pressures. Near their highest pressure, ~15 GPa, they observed that α-Fe transforms to a new phase, which they identified as hcp, now known as ε−hcp Fe (Jamieson and Lawson, 1962).

Jamieson and Lawson based their conclusion about ε−Fe on the observation of only one strong xrd line that was atypical of the bcc α phase of Fe and the constraint that their high-pressure phase should have the density of the high-pressure phase measured by Bancroft et al. under shock compression. That is, a second xrd line of the high-pressure phase of hcp-Fe could not be measured. So ε−Fe was assumed to be bcc, fcc or hcp, and a hypothetical second Fe xrd line for each possible phase was indexed such that the combination of the single measured xrd line and one of the possible bcc, fcc, or hcp lines would give the measured shock volume. Thus, ε-Fe was determined to be hcp.

The xrd measurements were a tour de force. The combination of the shock and static data resolved the question of α−ε transition in Fe at 13 GPa. Consistency of the static and shock experiments implied that the Fe $\alpha-\varepsilon$ transition had occurred under shock compression on a sub-μsec timescale. This agreement led to acceptance by the static-pressure community of the results of the shock experiment of Bancroft et al., which had not generally been accepted prior to the experiment of Jamieson and Lawson (Graham, 1994).

Bridgman's questioning of the interpretation of the shock-wave data of Bancroft et al. had set off a scientific controversy that motivated construction of a high-pressure system coupled to an X-ray diffraction system. It took five years to build the experimental apparatus and make the xrd measurements. The xrd data of Jamieson and Lawson and the resistance measurements of Drickhamer and Balkan resolved the question of whether a phase transition had occurred in Fe. Shock and static researchers have worked together on issues of pressure calibration ever since.

Once Bridgman realized that chemical explosives could be used reliably to achieve high-shock pressures, he shortly thereafter realized that nuclear explosives could be used to achieve even higher pressures. On the last page of the seventh and last volume of his collected works Bridgman wrote: "The very highest pressures will doubtless continue to be reached by some sort of shock-wave technique. ... Perhaps some fortunate experimenters may ultimately be able to command the use of atomic explosives in studying this field" (Bridgman, 1963; Bridgman, 1964, Vol. 7: no. 199–4625/4637).

4.8 Altshuler: The 1960s

Bridgman's suggestion about nuclear explosives was eventually read and pursued by L. V. Altshuler and his group at Arzamas-16. In his memoirs Altshuler (2001) writes: "In 1962 Bridgman ... suggested that with some luck, experimenters might even employ atomic blasts in high-pressure research. Such lucky experimenters were my team's members ... who in 1968 were the first to carry out measurements in the near zone of an underground nuclear explosion" (Altshuler et al., 1968). Today, shock experiments at ultrahigh pressures are done with giant pulsed lasers and with the giant pulsed-power Z machine, which drives metal impactors to ultrahigh velocities and impact-shock pressures.

4.9 A New Beginning

In the same paper Bridgman states: "It is conceivable that a way will be found of superimposing shock-wave pressures with static pressures." This kind of experiment is done today by pre-compressing a sample statically in a diamond anvil cell (DAC) and then shock compressing it to high pressure generated with a pulsed giant laser. Water, for example, has been pre-compressed to ~1 GPa in a DAC and then shock-compressed with a laser to 250 GPa (Lee et al., 2006). A variation of the above-mentioned pre-compression experiment is a reverberating shock-wave experiment in which the first shock effectively pre-compresses a liquid to a relatively low shock pressure and the remaining eight reverberating shocks then essentially compress the fluid isentropically from that first-shock state. Virtually, the same states are achieved in water up to 100 GPa using both methods (Chau et al., 2001; Lee et al., 2006).

To complete the Fe story, in the early 1970s Barker and Hollenbach (1974) developed the VISAR, a fast optical velocimeter with which they measured with ns time resolution the continuous temporal profile of a three-wave shock in Fe. Their measurements confirmed the three-wave structure in Fe, which had been observed by Bancroft et al. While Barker and Hollenbach confirmed a phase transition does occur, their technique is not able to determine the crystal structure of the high-pressure phase.

Still to be demonstrated experimentally was the structure of the high-pressure phase and the fact that the $\alpha-\varepsilon$ phase transition actually occurs on a sub-μsec timescale. In 2005, a laser beam was split into two beams, one to generate an intense 1-ns X-ray source and one to generate a shock in a small, oriented single crystal of Fe. Those results showed that an Fe phase transition occurs in ~1 ns under shock compression at 13 GPa and the in situ structure of the high-pressure phase is ε-hcp (Kalantar et al., 2005).

Bridgman was truly remarkable in terms of his contributions to static high-pressure research over his fifty-six-year career, the students he produced, his important contributions to establishing the scientific foundations of shock-compression research and his predictions about shock-compression research that have been demonstrated to be correct decades after his death.

High-pressure research remains a field that is primarily a curiosity-driven quest for discovery. Discovery-driven research is about ideas – an unexpected experimental observation that requires theoretical interpretation or a theoretical prediction that requires experimental verification. This iteration between experiment and theory is the foundation of science. It is so basic, in fact, that this process has been given a name: the scientific method. This principal is the essential element of all scientific and technological research and development.

Technological applications follow scientific discovery. Since scientific discoveries are generally unexpected, so too are their associated technological applications. Perhaps the greatest example of this process is the experimental discovery of the nuclear atom. In 1911 Ernest Rutherford demonstrated that atoms are composed of tiny nuclei with virtually all the atomic mass, surrounded by clouds of tiny, virtually mass-less electrons that occupy virtually all the atomic volume. When reportedly asked if his discovery had any practical applications, Rutherford is reported to have replied that there were none that he could think of. The nuclear atom, discovered as a result of pure intellectual curiosity, is the basis of the majority of high-technology developments, which today are the bases of the world's largest economies.

5

Rare Gas Fluids

Rare-gas liquids (RGLs) are one of the simplest types of condensed matter under dynamic compression. At ambient, RGLs are composed of atoms with filled electronic shells, which interact relatively weakly via van der Waals forces between atom pairs. As a result these liquids are quite compressible, and shock-induced temperatures are substantial. Because liquids are disordered, shock-induced entropies of RGLs are relatively small compared to those of shock-compressed crystals. Thus, shock-induced dissipation of rare-gas fluids (RGFs) is primarily shock heating (Nellis, 2005, 2006a).

Because pair interactions are relatively weak, the lowest lying electronic excitation energies of liquid He, Ne, Ar, Kr and Xe are nearly equal to the ionization energies of the respective rare-gas atoms. Electronic band gaps E_{gap} between filled valence-electron states and empty conduction-electron states are 9–25 eV for Xe through He (Rossler, 1976; Schmidt and Illenberger, 2005). For temperatures $T < 290$ K (0.025 eV), the gas–liquid critical temperature of Xe, which is the largest of the gas–liquid critical temperatures of RGLs (Vargaftik, 1975), is such that $T << T_{gap} = E_{gap}/k_B$, where k_B is Boltzmann's constant. Thus, dense RGFs interact via weak effective van der Waals-type pair potentials and are (1) electrical insulators at temperatures T less than $\approx 0.1\ T_{gap}$ and (2) semiconductors at temperatures T greater than $\approx 0.1\ T_{gap}$. Wigner-Seitz radius R_{WS} is the radius of a spherical atom with volume corresponding to molar volume $V_0 = N_0\ (4\pi/3)(R_{WS})^3$, where N_0 is Avagadro's number and $V_o = 1/\rho_0$. Parameters of RGLs at atmospheric pressure are given in Table 5.1.

Melting temperatures of solid Ar, Kr and Xe at 100 GPa static pressures range up to ~3000 K (Boehler et al., 2001). These temperatures are very large compared to their melting temperatures of ~100 K near atmospheric pressures. However, shock temperatures of RGFs at 100 GPa shock pressures are greater than 10,000 K (Nellis, 2005).

Table 5.1 *Boiling temperatures* T_0, *mass densities* ρ_0 *(g/cm*^{3}*), number densities* n_0 *(*10^{22}*/cm*^{3}*), first ionization potentials* E_1 *of rare-gas liquids at atmospheric pressure and Wigner-Seitz radii of rare-gas atoms (Nellis, 2005).*

	T_0(K)	ρ_0 (g/cm^3)	n_0 (10^{22}/cm^3)	E_1, (eV)	R_{ws}(Å)
He	4.2	0.125	1.9	25.5	2.33
Ne	27.1	1.21	3.6	21.6	1.87
Ar	87.3	1.40	2.1	14.4	2.25
Kr	119.8	2.42	1.7	11.6	2.40
Xe	165.2	2.96	1.4	9.2	2.58

5.1 Single-Shock Compression

At shock pressures below ~50 GPa, the dominant interaction in supercritical RGFs is an effective exponential-six pair potential:

$$\phi(r) = [\varepsilon/(\alpha - 6)]\left\{6\exp\left(\alpha[1 - r/r^*] - \alpha(r^*/r)\right)^6\right\}, \tag{5a}$$

$$\varepsilon = (k_B T_c)/1.2375, \tag{5b}$$

$$r^* = \{18.1896(T_c/P_c)\}^{1/3}, \tag{5c}$$

$$\alpha = 13.0, \tag{5d}$$

where T_c and P_c are critical temperature and pressure, respectively (Ross and Ree, 1980). The parameters of this potential are listed for monatomic RGFs in Table 5.2. The depth of the well between the repulsive and attractive portions of this potential ranges between 4 K and 230 K, which means that the repulsive portion of the potential dominates shock-compressed RGFs whose shock temperatures herein are generally well above 1000 K.

At shock temperatures $T_H > 0.1T_{gap}$, which means shock pressures above ~50 GPa, the dominant interactions are an effective inter-atomic exponential-six pair potential plus electronic excitations across E_{gap}. Because T_H increases with shock pressure and E_{gap} decreases with shock density, the combination generates substantial electronic excitation and associated changes in physical properties, such as increasing electrical conductivity with increasing shock pressure. For example, dense fluid Ar has been shock-compressed to 91 GPa and nearly threefold liquid density. At those conditions a sufficiently high density of thermal electrons is excited to enable an estimate of $E_{gap}(\rho)$ at high densities (Ross et al., 1979). This idea was later used to observe the crossover from semiconducting fluid H to MFH (Weir et al., 1996; Chapter 6).

At sufficiently high densities and temperatures, RGFs are poor metals with minimum metallic conductivity (MMC) for which the mean-free path of scattering

Table 5.2 *Parameters of exp-6 potentials in Eqs. (5a)–(5c) for rare-gas fluids at high shock pressures and temperatures derived from corresponding-states scaling (Ross and Ree, 1980).*

	ε/k_B (K)	r^*(Å)	α
He	4.2	3.48	13.0
Ne	36	3.15	13.0
Ar	122	3.85	13.0
Kr	170	4.14	13.0
Xe	234	4.49	13.0

of delocalized thermally activated electrons is comparable to average inter-atomic distance (Ioffe and Regal, 1960; Mott, 1972). Elemental RGFs have similar properties because of their similar atomic-fluid structures. For this reason, in this work we focus primarily on Ar. Results for the other RGFs are similar and are discussed elsewhere (Nellis, 2005).

Smoothed measured Hugoniot data of liquid Ar (van Thiel and Alder, 1966b; Nellis and Mitchell, 1980a; Grigoryev et al., 1985) indicate substantial thermal electronic excitation above 50 GPa, as will be discussed. Total specific shock energy E_H is represented by the area of the triangle under the Rayleigh line. Internal energy caused by isentropic compression is reversible by definition. Because the isentrope of Ar is essentially coincident with its 300-K isotherm, the reversible energy of Ar is essentially the area under its isotherm. For a shock pressure of 73 GPa in Ar, substantially greater irreversible energy is deposited than reversible energy, ~90% and ~10%, respectively (Nellis, 2005, 2006a).

At $P_H = 50$ GPa and $V_H = 13$ cm^3/mol, T_H ~1.2 eV (Grigoryev et al., 1985), at which sufficient electronic excitation occurs to affect the P_H-V_H slope of the Hugoniot curve. At that point, $T_H \sim T_{gap}/10$, which is a common criterion for appearance of effects of electronic excitation on Hugoniots of RGFs. At shock pressures above 50 GPa in Ar, internal energy that otherwise would go into temperature and associated thermal pressure is instead absorbed internally by thermal excitation of electrons. Irreversible internal thermal energy is commonly known as shock heating and is the reason shock pressure is higher than static pressure at compressions to the same given volume.

Bulk modulus $B = -V(dP/dV)_V$ is the reciprocal of compressibility and is a measure the pressure required to achieve a fractional decrease in volume. B generally increases with increasing pressure and does so whether compression is isothermal, isentropic or adiabatic (shock compression). For example, a shock pressure of 5 GPa is required to compress liquid Ar by a factor of 1.6 in density (van Thiel and Alder, 1966b). A shock pressure of 100 GPa compresses liquid Ar by a factor of 2.9. While 5 GPa causes a factor of 1.6 increase in density, to increase fluid Ar density an additional 1.3-fold requires a much larger shock pressure of 100 GPa. An adiabatic compression of 2.9-fold at 100 GPa is less than a factor of 4, the limiting shock compression of both an ideal monatomic gas and a degenerate free-electron gas (Nellis, 2003).

Electrical conductivities and brightness temperatures have been measured on the Hugoniots of Ar, Kr and Xe up to 90 GPa (Mochalov et al., 2000). Spectral emissions were measured at 430, 500 and 600 nm and temperatures were obtained by fitting those spectral intensities to gray-body spectra. The conductivities have exponential dependences on 1/T and, in the cases of Kr and Xe, reach plateaus of ~600/(Ω-cm) at the highest shock pressures. This value is essentially MMC of a disordered metal, typically ~1000/(Ω-cm), as expected in disordered metallic fluids (Mott, 1972; Collins et al., 1998).

Electrical conductivities of liquid Ar have been measured under shock pressures in the range from 20 to 70 GPa. The effective activation energy of Ar derived from a plot of the logarithm of conductivity versus $(k_BT)^{-1}$ is 14.8 eV, which is large compared to the highest shock temperature achieved of ~1.5 eV. The maximum measured conductivity is ~100/(Ω-cm) at 70 GPa, which indicates that Ar is semiconducting at 70 GPa on the Hugoniot and probably approaches MMC at higher densities (Gatilov et al., 1985). Thermally activated fluid Ar was treated under the assumption that thermal activation energy is independent of density (Mochalov et al., 2000), even though shock densities vary by ~50% along this Hugoniot. However, this method gives reasonable estimates.

Experimental and theoretical results for electrical conductivities of He, Ne, Ar, Kr and Xe shock-compressed with chemical explosive to 100 GPa pressures have been achieved by shock and quasi-isentropic compression. At highest dynamic pressures, electrical conductivities extend into the range from 10^2 to 10^3/(Ω-cm), typical of MMC (Fortov et al., 2003). Thermal conductivities and thermopowers of Xe have been investigated. Theoretical calculations used a partially ionized plasma model (Fortov et al., 2003; Kuhlbrodt et al., 2005; Mintsev and Fortov, 2015).

The shock compression curves of liquid Xe and Kr have been measured at the Z Accelerator up to 850 GPa at which $10 < u_p < 16$ km/s (Root et al., 2010; Mattesson et al., 2014). Calculated shock temperatures range up to 150,000 K for both. In this range of shock pressures $u_s = C + Su_p$, where C and S are constants. For Xe, C = 1.624 km/s and S = 1.163; for Kr, C = 1.313 km/s and S = 1.231. The S parameters of Xe and Kr are ~1.2, as is systematically the case at TPa shock pressures for many dense fluid metals (Ozaki et al., 2016). The C and S parameters of the Hugoniots of Xe and Kr are ~1.5 km/s and 1.2, respectively, which is typical of compressible fluid D_2 at shock pressures of ~100 GPa (Boriskov et al., 2005).

He has been shock-compressed to as much as twelvefold liquid He density at 100 GPa pressures. Highest compressions were achieved by pre-compressing liquid He at room temperature a factor of 3.3 in a DAC. Those samples were then shock-compressed fourfold with a giant pulsed laser. Compression increases at the onset of ionization (Eggert et al., 2008), analogous to the increase in compression in Ar at the onset of electronic thermal excitations at ~50 GPa.

5.2 Quasi-Isentropic Compression in Converging Cylindrical Geometry

Liquid argon has been compressed quasi-isentropically by a converging metallic cylinder (Urlin et al., 1997). The volume of the compressed Ar was measured with flash X-radiography, and the pressure was calculated by computationally simulating the hydrodynamics. Those experiments achieved Ar densities of 4.8, 6.1 and 7.3 g/cm^3 at calculated pressures of 93, 247 and 480 GPa and calculated temperatures of 6,600, 11,400 and 19,000 K, respectively.

5.3 Multiple-Shock Compression

Compressed Ar gas initially at a pressure of about 20 MPa has been multiply shock-compressed with a reverberating shock wave. Shock waves were generated by impact of projectiles accelerated to a velocity as high as 6 km/s. Time-resolved radiated emission histories were measured with an eight-channel pyrometer operating at wavelengths between 400 and 800 nm. Time-resolved particle velocity histories were measured with a Doppler pin system (DPS). Hugoniot, double-shock and quadruple-shock states were measured at peak pressures up to 21, 73 and 158 GPa, respectively. The results were used to verify self-consistent variational theoretical models with partial ionization (Chen et al., 2014). Similar experiments have been performed on dense fluid neon and krypton (Chen et al., 2015).

6

Metallization of Fluid Hydrogen at 140 GPa

Hydrogen is ubiquitous in the universe with a relative elemental solar abundance of ~0.92 (Arnett, 1996). The corresponding abundances of He and the total of everything else are ~0.07 and ~0.01, respectively. Naturally occurring condensed hydrogen in planets is primarily in the liquid or fluid states within our solar system. The H atom is composed of one electron and one proton, the simplest of all atoms. In ultracondensed form at ninefold H density in liquid H_2 at 20 K, Wigner and Huntington (1935) predicted that solid H_2 would undergo an IMT at 0.62 mol H/cm^3 at "very low temperatures" at a pressure greater than 25 GPa.

Hydrogen in the form of its isotopes deuterium (D) and tritium (T) would solve the world's energy needs for centuries if D and T could be made to efficiently undergo nuclear fusion (Feynman, 1963) by ICF at extremely high densities and temperatures obtained by laser-driven dynamic compression or by nuclear fusion of relatively low-density D-T plasma contained by a magnetic field. If metallic H made at extremely high pressures and densities could be retained metastably on release to ambient pressure, then metallic solid H (MSH) has the potential to revolutionize life as we know it in the forms of high-density D-T fuel for ICF; fuels and propellants for commercial vehicles and rockets for space travel, respectively; potentially light-weight structural materials; and for unusual quantum materials at room temperatures with potentially unusual material properties, such as superconductivity (Nellis, 1999). In the early 1990s a metallic phase of dense H had yet to be made, despite its predicted existence sixty years previous by WH and its many attractive potential technological applications and scientific challenges.

The first key scientific challenge presented by dense hydrogen was the challenge to make ultracondensed metallic hydrogen in any phase. Solid H_2 had been compressed under static compression in DAC to densities well above the density at which WH had predicted electrically insulating H_2 would transition to metallic H.

At WH's predicted density of 0.62 mol H/cm^3 = 0.31 mol H_2 /cm^3 and 300 K, we now know that expected static pressure is 73 GPa (Loubeyre et al., 1996). In the 1990s pressures on solid H_2 in DACs were up to ~200 GPa and higher. Thus, it appeared that metallic H might be made simply by dissociation of H_2 to H at above ~100 GPa pressures, if only highly stable dense H_2 would dissociate to H.

Shock compression generates temperature T and entropy S in the thin front of shock wave. Phase stability is determined by T and S. That is, Helmholtz free energy $F = U - TS$, where U is internal energy. Thus, again it appeared likely that if dense H_2 could be made to dissociate to dense H at dynamic pressures available in a laboratory, metallic H might be made simply as a consequence of dissociation. So we designed a sample holder specifically to make metallic H and conducted a series of multiple-shock compression experiments, which generate T and S. In order to detect metallization, electrical conductivities were measured directly as a function of dynamic P, ρ and T.

Because finite Ts are generated at increasing ρs under quasi-isentropic (QI) multiple-shock compression, the energy gap $E_g(\rho)$ between filled valence states and empty conduction states of hydrogen decreases with increasing dynamic pressures and densities, as for shock compressed fluid Ar (Ross et al., 1979). In this picture measured electrical conductivity of hydrogen is expected to increase exponentially until $E_g(\rho)$ is filled in with thermally excited elections, which is expected when $E_g(\rho_{met}) \approx k_b T_{met}$ at metallzation density ρ_{met}.

In those dynamic-compression experiments MFH was made at the density ρ and the general conditions of pressure P and temperature T predicted by WH (1935). Observed metallization density is ninefold liquid H density at 20 K. MFH was achieved with multiple-shock-induced dissipation T and S at sufficiently high P and ρ and sufficiently low T to make a degenerate metal by overlap of $1s^1$ electronic wave functions on adjacent H atoms (Weir et al., 1996a; Nellis et al., 1999; Fortov et al., 2003; Nellis, 2013, 2015a).

However, traditional single-shock compression would not be sufficient. Achieving desired thermodynamics to make a metal required tuning the hydrodynamics of dynamic compression to obtain appropriate ρ and T, which is readily achieved. For example, because number densities of liquid H_2 and D_2 at 20 K and 1 bar differ by a factor of 2.4 (rather than 2.0), various values ρ and T are readily obtained at a various final pressures P.

Dynamic compression was used specifically so that temperature and entropy generated by the multiple-shock process would dissociate H_2 molecules and so drive a crossover at finite T to a poor metal. Metallization of H is observed at P_{met} = 140 GPa with density ρ_{met} and temperature T_{met} such that $E_{gap}(\rho_{met})\approx k_B T_{met}$, where $E_{gap}(\rho)$ is mobility gap as a function of density and k_B is Boltzmann's constant (Weir et al., 1996a; Nellis et al., 1999).

At H metallization ρ_{met} and T_{met}, degeneracy factor $\eta = T/T_F \approx 0.014 << 1$, where T_F is Fermi T. Wigner-Seitz metallization radius $r_s = 1.63$. MFH is probably monatomic based on density functional calculations (Pfaffenzeller and Hohl, 1997), on the high temperatures of 1700 to 3000 K and very high densities of the crossover in fluid H from semiconducting to MMC (Nellis et al., 1999), as well as systematic correlations of measured electrical conductivities with radial density dependences of atomic wave functions of H, O, N, Rb and Cs (Chau et al., 2003a). Temperatures at those measured MMCs exceed 2000 K at high densities, at which diatomic molecules are expected to dissociate. Highest melting temperature observed to date of solid H_2 is 1,060 K at 65 GPa (Deemyad and Silvera, 2008).

Initial liquid H_2 sample diameters and thicknesses were 25 mm and 0.5 mm, respectively, which facilitated electrical conductivity measurements with electrical leads during experimental lifetimes of ~100 ns at dynamic pressure from 90 to 180 GPa (1.8 Mbar) and temperatures up to 3000 K (Nellis, 2000). Because of the finite temperatures and large ~15 eV electronic band gap of H_2 at ambient pressure (Inoue et al., 1979), the crossover from semiconducting to metallic fluid H could be tracked by measurement of the density and temperature dependence of electrical conductivity. Shock-induced dissipation T and S in a tuned QI pressure pulse drove diatomic dissociation, which enabled discovery of MFH.

6.1 A Little History

WH made their classic prediction of metallic solid hydrogen at density 0.62 mol H/cm^3, "very low temperatures" and a pressure greater than 25 GPa (Wigner and Huntington, 1935; Nellis, 2013). Their predicted metallization density is ninefold liquid-hydrogen density at 20 K and was based simply on total energy calculations of H_2 and H on a bcc crystal lattice. The fact that hydrogen has never been observed in a bcc crystal lattice is not of great importance. Electron band theory had yet to be developed at the time of that prediction. In a condensed matter sense T need not be absolutely low. That is, temperatures of a metal are low if the Fermi-Dirac electron distribution is highly degenerate, that is, $T/T_F << 1$, where T_F is Fermi temperature (Mott, 1936).

In 1956 when a static pressure of 2 GPa was first achieved in H_2 (Stewart, 1956; Anderson and Swenson, 1974), making metallic hydrogen became a prime goal of the static high-pressure community. In the late 1970s, 100 GPa static pressures were reached in a DAC. Inoue et al. (1979) measured the 15 eV electronic band gap (E_g) of solid H_2. In the 1980s, research on metallization of solid hydrogen was based on the assumption that at sufficiently high static pressure, the 15-eV band gap of condensed H_2 at atmospheric pressure would close to $E_g \approx 0$, and thus hydrogen would eventually become a metal at some high pressure P_{met} and density

ρ_{met}, which would probably be achieved experimentally in a DAC. However, by the early 1990s static high-pressure experiments had demonstrated that solid insulating H_2 neither dissociates nor metallizes up to the highest static pressures achieved at that time (~300 GPa) and the required metallization pressure near $T = 0$ was unknown.

At WH's predicted metallization density ρ_{met} = 0.62 mol H/cm^3 = 0.31 mol H_2/cm^3, static pressure is 73 GPa, based on the 300-K P-ρ standard of Loubeyre et al. (1996). At 73 GPa the H_2 electronic band gap in the solid is ~8 eV (Mao and Hemley, 1994), definitely not a metal. Near 73 GPa, solid H_2 is observed experimentally to be an insulator in a complex regime with variations of molecular orientations, bond lengths and *c/a* ratios (Kaxiras et al., 1991).

Ashcroft (1991) said that "the geometric and dynamic nature of the (H-H) pairing are both crucial to the preservation of the insulating state up to the very substantial densities where its continued persistence has been experimentally confirmed." WH themselves were not certain about their classic IMT prediction in hydrogen. WH said that if their classical predicted IMT does not occur, then a transition to a different structure, perhaps with a laminar crystal structure, similar to hexagonal close packed (hcp), might occur at some high pressure. By the mid 1990s the static pressure corresponding to the predicted density of WH's famous IMT had been surpassed in laboratories by a factor of 3–4 without observation of the predicted metallization transition. Despite "heroic" experimental efforts, the H_2 proton-proton bond was yet to be broken under P alone at 300 K, and metallic solid hydrogen was still yet to be made at static high pressures achieved to date, at that time up to ~400 GPa in a DAC (Dalladay-Simpson et al., 2016).

All these considerations imply that if diatomic H_2 could be driven to dissociate to monatomic H at high pressures, then pressures already achieved on hydrogen might be sufficiently large to make metallic H. The primary purpose of this chapter is to discuss the general philosophy of the sample holder designed specifically to make MFH at a pressure that could be achieved in a laboratory at that time. That design is a combination of condensed matter physics and chemistry integrated with mechanical aspects of supersonic compression to drive a bonds-to-band transition.

6.2 What to Try?

In the early 1990s it was clear that another technique should be tried in the hunt for metallic hydrogen, in addition to static compression in a DAC at ~300 K. One logical technique to try would be to focus on driving dissociation from dense H_2 to H at already achievable high pressures. In this context, H metallization is viewed simply as an issue of phase stability, which is determined by free energy F. Helmholtz free energy is given by

$$F = U - TS. \tag{6.1}$$

Dynamic compression generates dissipation T and S, which can be used to control phase stability and thus might be used to make metallic H. Further, cryogenic technology and fast diagnostics were in hand to make liquid H_2 samples at the muzzle of a hypervelocity two-stage light-gas gun and to make the required fast voltage measurements on ns timescales. As William Shakespeare wrote in *Julius Caesar*, "There is a tide in the affairs of men, which taken at the flood, leads on to fortune." It was time to do some dynamic experiments.

6.3 Dynamic Compression of Liquid Hydrogen

Shock-induced dissipation T and S occur instantly on dynamic compression within the shock front. Shock-front widths in a fluid are of order ~nm and ~ps (Fig. 2.2), which are negligible compared to sample spatial thickness and temporal lifetimes of experiments, in this case ~50 μm and ~100 ns, respectively. Because the width of a shock front is by definition the spatial/temporal intervals required to achieve thermal equilibrium, dynamically compressed fluid hydrogen equilibrates thermally in ~ps and ~nm. Time resolution of measured conductivity voltages was ~ns. Hydrogen samples in those conductivity measurements were in thermal equilibrium.

Liquid-H_2 and liquid-D_2 samples were 25 mm in diameter and 0.5 mm thick. Liquid-H_2 at atmospheric pressure was used as coolant in all experiments. Liquid H_2 and D_2 samples were condensed from their respective gases in a pre-cooled sample holder at 20 K. Liquid samples under gravity in a sample holder have uniform initial density ρ_0 and initial temperature T_0, which are obtained simply by the measurement of barometric pressure and published liquid saturation curves.

Because thermodynamic states achieved with adiabatic dynamic compression are tunable, a crucial issue is identification of an appropriate dynamic pressure pulse to generate appropriate thermodynamic states in situ. Extensive measurements of Hugoniot equations of state and single-shock temperatures of liquid D_2 and H_2 have been made (van Thiel and Alder, 1966a; van Thiel et al., 1974; Dick and Kerley, 1980; Nellis et al., 1983; Ross et al., 1983; Holmes et al., 1995; Knudson et al., 2001, 2003a, 2004; Belov et al., 2002; Boriskov et al., 2005; Hicks et al., 2009). Fluid D_2 dissociates to D at 100 GPa shock pressures (Nellis, 2002b; Boriskov et al., 2005). Measured shock temperatures of liquid H_2 increase faster with shock pressure than melting temperature T_M of solid H_2 increases with static P (Deemyad and Silvera, 2008). Thus, shock-compressed liquid H_2 was expected to remain in the fluid phase at high dynamic pressures in the ~100 GPa range.

Limiting (maximum) shock compression of an ideal gas is fourfold of initial density, which is much less than ninefold compressed liquid-H_2 density WH predicted for metallization (1935). Measured limiting shock compression of liquid D_2 is 4.3-fold at 100 GPa pressures (Boriskov et al., 2005), close to calculated limiting shock-compressions of 4.3-fold (Militzer and Ceperley, 2000) and more recently 4.85-fold (Tubman et al., 2015). Liquid H_2/D_2, N_2, O_2 and CO all approach fourfold compression asymptotically at high shock pressures (Nellis, 2002b). Thus, single-shock compression is incapable of making MFH in liquid samples because shock-induced T limits compressed density to a value too small (~ fourfold liquid D_2/H_2 density) to make metallic H by overlap of electron wave functions on adjacent H atoms at ninefold compressed liquid-H_2 density predicted by WH.

In contrast, multiple-shock compression of hydrogen is QI – virtually isentropic from the first-shock state (Fig. 2.8). With a 2SG, QI compression achieves ultra-condensed degenerate fluid H at high pressures up to~200 GPa. The pressure pulse applied to fluid hydrogen was achieved in situ by a sequence of more than eight relatively weak shock waves. Each of those weak shocks contributes some amount of T and S. For this reason the crossover at the high pressures and temperatures achieved under dynamic compression was expected to be a continuous crossover in dissociation from fluid H_2 to H with a corresponding crossover in electrical conductivity from semiconducting fluid H to poor-metallic fluid H with MMC.

Although WH predicted metallization density at "very low temperatures," thermal energies were neglected in their calculations. Thus, their predicted density is effectively at 0 K, near which phase transitions are generally first-order. In this case a discontinuous IMT was expected because of finite dynamic temperatures at predicted metallization density. In fact, a continuous crossover was observed in the electrical conductivity experiments.

The goal of the experimental design became one of identifying a dynamic pressure pulse that produces less heating than single-shock compression but sufficient T and S to achieve sufficiently high density and low temperature in degenerate fluid H to make MFH. Because liquid H_2 is a compressible van der Waals liquid, shock dissipation is absorbed substantially by large compressions with associated shock-heating T. Increased entropy S of liquid H_2 is shock-induced by dissociation of H_2 molecules.

QI compression is achieved adiabatically by shock reverberation in low-density liquid H_2 contained between two high-density sapphire anvils. Sapphire anvils achieve up to ~tenfold compression of liquid H_2 by more than ~eight relatively small shocks, which meant metallic hydrogen might actually be made in such experiments and so sapphire anvils were selected. Sapphire anvils were also used for a practical reason. At the time of the QI compression experiments, sapphire was the only dense, strong, electrically insulating material whose Hugoniot had been measured above 100 GPa and was thus suitable for such use.

The issue with anvil material stemmed from the fact that limited funding was available for the project, which meant it was not possible to perform a systematic investigation on a number of possible materials to identify an optimal anvil material. For this reason, sapphire was selected and performed ideally for the electrical conductivity experiments. However, sapphire becomes opaque by shock-induced defects at ~100 GPa pressures (Urtiew, 1974), which is why black-body thermal-emission temperatures of MFH have yet to be measured. A suitable transparent anvil material at shock pressures above ~100 GPa is needed but has yet to be identified.

6.3.1 Condensed H_2 and D_2 Samples

Hydrogen and deuterium samples were used to tune densities and temperatures achieved under high dynamic compression because of their significantly different number densities and corresponding different effects on the hydrodynamics. Liquid H_2 at 20 K was used as a coolant for experiments using both liquid H_2 and liquid D_2 samples. Because temperatures achieved under dynamic compression range from 1700 to 3000 K, there is no quantum H/D isotope effect at those high shock pressures. However, there is an H/D density isotope effect in H_2 and D_2 samples at 20 K. The ratio of density of liquid D_2 to H_2 at 20 K is 2.4, greater than 2 because of greater zero-point motion in lighter H. At 20 K liquid D_2 has a higher number density than liquid H_2, which causes different mass densities and temperatures at a given high dynamic pressures, a fact used to tune thermodynamics at extreme conditions.

High dynamic pressures were generated by impactors accelerated to velocities as large as 7 km/s with a 2SG. Samples were contained in cryogenic sample holders at 20 K. Liquid H_2 and D_2 samples were at a stable temperature near the normal boiling point of liquid H_2 at $T_0 = 20$ K near 1 bar at the time the gun fired an impactor. Because this cryogenic system had been used successfully many times to single-shock compress cryogenic liquids, the sample holder modification for multiple-shock compression required only the simple insertion of two sapphire anvil disks.

To minimize dynamic temperature to achieve ultracondensed degenerate matter, it is necessary to maximize initial sample density. Sample density of a gas is maximized in two ways. First, condensing H_2 or D_2 gas to a liquid requires a cryogenic sample holder, which we had developed previously (Nellis and Mitchell, 1980; Nellis et al., 1983). By condensing to a liquid rather than a solid, one is certain that liquid under gravity completely fills the sample volume with no imperfections such as voids or cracks as is possible in a condensed solid. Liquid H_2 and D_2 samples were placed under H_2 or D_2 gas pressure of ~860 Torr, respectively, to suppress bubbles caused by boiling to maintain planar shock fronts with minimum distortions caused by bubbles. The cryogenic system is described (Nellis et al., 1999), and photographs of the 2SG system and cryogenic sample holder are published (Nellis, 2007a).

A second option to maximize sample density was a static pressure system to apply ~0.1 GPa to gas in a compressed-gas sample holder, which would have required special material to withstand the applied gas pressure with minimum distortions of the sample holder and a static-pressure gas system would need to be developed. Hugoniots of any materials needed in data analysis would have to be measured. The choice of the existing cryogenic system used for single-shock experiments was obvious.

6.3.2 Tuning Thermodynamics

Desired thermodynamics were achieved by tuning supersonic hydrodynamics to generate QI compression, which in this instance meant increasing the rise time of pressure to ~10 – 100 ns (Nellis, 2006a) relative to the fast rise time of a single-shock wave in a liquid (< 1 ns). Two ways are available to increase rise time of pressure: multiple-shock and ramp-wave compressions. The former is a sequence of discrete step increases and the latter is continuous increase of pressure with time. Equivalent states are achieved with both. Multiple shocks are readily achieved with hypervelocity impact, our chosen method, because it was readily achievable with existing facilities. Pressure pulses were generated by hypervelocity impact of metal plates accelerated with a 2SG to velocities as high as ~7 km/s onto cryogenic sample holders. Multiple shocks are generated in situ simply by shock reverberation between two anvils. Impactor velocity was measured with an existing flash X-ray system with an accuracy of 0.1% (Mitchell and Nellis, 1981a).

For an ideal gas, Fig. 2.8 shows that a sufficiently large number of multiple shocks to achieve QI compression consists of an initial weak shock H_1 plus a sequence of successive multiple shocks on the isentrope starting from that first-shock state H_1. Multiple shocks were calculated with repeated use of the Rankine-Hugoniot equations, assuming various incremental increases in shock density. Shock density increments are caused by shock reflections in liquid off weakly compressible anvils. An isentrop is the limit of an infinite number of infinitely weak shock waves (Zeldovich and Raizer, 1966). The maximum number of multiple shocks possible on reverberation is achieved with the largest shock-impedance mismatch between liquid H_2 (0.07 g/cm^3) and the two sapphire anvils (4.00 g/cm^3). Shock-impedance, $\rho_0 u_s$, of an anvil is maximized by maximum density. Sapphire was chosen for the anvil material for its high density and because the Hugoniot of sapphire had already been measured (Marsh, 1980; Erskine, 1994).

In this case, electrical leads would be inserted through an anvil to contact conducting fluid hydrogen, which means an anvil could not be a conductor under shock compression. Electrical conductivities of anvil materials under high shock pressures needed to be measured to verify that their shock-induced conductivities are sufficiently low. Transparent insulators are desired to permit thermal radiation to exit compressed and heated fluid hydrogen to measure its temperature by blackbody emission, if possible. The Hugoniot of the anvils is required in data analysis. Strong transparent oxides are weakly compressible, which means shock dissipation in an anvil is mainly by entropy and thus shock temperatures and thermal pressures are relatively close to those of isentropes, which means that such a material is ideal for designing shock reverberation experiments. Also, because two anvils are destroyed in each experiment, their cost is an issue. The only known material at the time that comes close to matching these requirements is sapphire, single-crystal Al_2O_3.

Two sapphire disks 25 mm in diameter and 2.0 mm thick were used as anvils in shock reverberation experiments. Metal electrical contacts inserted through the sapphire disks were made of stainless steel, which is reasonably shock-impedance matched to sapphire and has relatively low electrical conductivity. The latter is needed for fast magnetic flux diffusion to establish uniform electrical current density over cross-sectional areas of the electrodes inserted through an anvil. The latter is a boundary condition on the iterative solution of the finite-element calculations of the cell constant.

Electrical resistance R of conducting fluid H and D were measured with fast oscilloscopes. To derive electrical conductivity σ from measured R, the cell constant C_i must be known, where $R = C_i/\sigma$. C_i is also known as the geometrical factor because R and σ are related by the shape of the conductor. C_i was calculated with three-dimensional (3D) computer simulations of current flow in the sample geometry, which was a disc thin compared to its outer dimension. The sample was simulated with a finely meshed 3D cubic network of resistors. By using Kirchoff's law that net current flow into any node is zero, a self-consistent solution for current flow in the two-probe network was obtained with an iterative method. Cell constant C_2 was found analogously for a four-probe network.

Calculated cell constants were checked experimentally on a tabletop with a mock cell using an electrolytic solution whose thickness could be varied accurately between 0.2 and 1.0 mm. Because of good agreement between calculated and measured cell constants for thicknesses down to 0.2 mm, cell constants were calculated for sample thicknesses in the conductivity experiments under dynamic compression, which were ~0.05 mm (50 μm) calculated with hydrodynamic simulations. Sample resistances were large compared to contact resistances. Thus, the two- and four-probe contact geometries yielded similar sample conductivities.

Experiments to measure electrical conductivities of dense fluid hydrogen are illustrated in Fig. 2.16. Fig. 1.2 illustrates schematically pressure histories for single-shock and multiple-shock compressions, both to the same final pressure. Fig. 1.3 illustrates schematically the respective pressure-density curves that are achieved by the two pressure histories in Fig. 1.2. MMC is achieved at 140 GPa in the sample holder in Fig. 2.16. The calculated planar pressure history at the longitudinal midpoint on the central cylindrical axis of the hydrogen layer is plotted in Fig. 2.17. That pressure history is generated by the shock wave initiated by impact onto the sample holder, which shock then reverberates in fluid H_2 back and forth between two sapphire anvils.

P-u_p are the only shock variables that are continuous across a boundary between two materials. Two anvils are used to calculate wave reflections by shock-impedance matching at the two hydrogen-sapphire interfaces. When a shock wave travelling though hydrogen impacts a sapphire wall, the shock formed in hydrogen on reflection is determined to a good approximation by assuming that re-shock curves of hydrogen are mirror reflections in P-u_p space of the Hugoniot of hydrogen. The ~100 ns experimental lifetimes are more than sufficiently long to achieve thermal equilibrium in fluid H, while being sufficiently brief to prevent significant hydrogen from diffusing out of its holder into the surrounding walls.

A key requirement of the sample holder is that it facilitates maximum sensitivity and resolution in measurements of the expected values of electrical conductivities. This is a very important issue because conductivities to be measured range over several orders of magnitude. In experiments discussed herein sample electrical resistance was determined as voltage drop across the sample measured by fast oscilloscopes divided by electrical current through the sample. Sensitivity of the oscilloscopes to measure voltages in a wide range extending over several orders of magnitude of potential values was maximized by inserting a resistor of known electrical resistance in parallel with the sample resistance and measuring the resistance of the parallel combination. The resistor added in parallel with the sample resistance is called a shunt resistor. The expected value of sample resistance to be measured determines the optimal value of shunt resistance used in a given experiment. For the electrical conductivities of dense fluid H measured, the shunt resistor varied from ~infinite (no shunt resistor) for very low H conductivities to a ~1 ohm shunt resistor for conductivities of MFH (Nellis et al., 1999).

6.3.3 Electrical Conductivities of Dense Fluid H/D from 90 to 180 GPa

Measured electrical resistivities of dense fluid hydrogen and deuterium at pressures P_f in the range 90 to 180 GPa are plotted in Fig. 6.1. Pressures on the abscissa are maximum "ring-up" pressure (P_f in Fig. 2.17). Values of P_f were the peak initial pressure in Al_2O_3 generated by impact in each of ten experiments. P_f was determined experimentally by shock impedance matching each of the ten experiments using measured impact velocities in each experiment and previously measured Hugoniot equations of state of Cu, Al and Al_2O_3, as appropriate. P_f is also the peak dynamic pressure generated in hydrogen in each experiment.

In the shock-reverberation experiment calculated in Fig. 2.17, the impactor was a Cu plate at measured impact velocity $u_I = 5.58$ km/s, which means $P_f = 140$ GPa in both fluid H and in the sapphire anvils by shock-impedance matching. Metallization of H was observed at 140 GPa, density ρ_{met} and temperature T_{met} at which $E_{gap}(\rho_{met}) \approx k_B T_{met}$, where $E_{gap}(\rho)$ is mobility gap of H as a function of density and k_B is Boltzmann's constant (Weir et al., 1996a; Nellis et al., 1999).

For the lower values of P_f, values of T_f using liquid H_2 samples were so small that conductivities were too small to measure accurately. Thus, for some lower values of P_f, liquid D_2 samples were used because the shock-impedance match

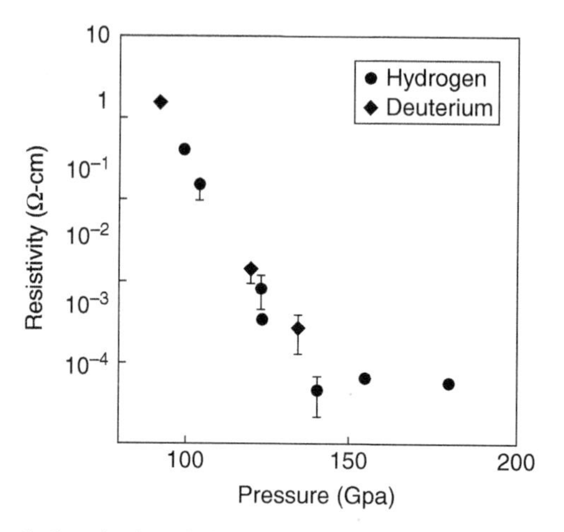

Fig. 6.1. Measured electrical resistivities of fluid D and H versus multiple-shock pressures P_f determined experimentally by shock-impedance matching. At the high temperatures above 1500 K achieved at high pressures in these experiments, quantum-isotopic effects in atomic H and D are negligible. However, at 20 K initial sample temperatures, quantum-isotopic effects are present, which cause different initial number densities and thus different final temperatures and densities for liquid H_2 and D_2 samples. In this way, liquid H_2 and D_2 samples were used to tune final thermodynamic states achieved in fluid H and D. Copyright 1999 by American Physical Society.

with a liquid D_2 sample generates higher temperatures and thus electrical conductivities in liquid D_2 than for H_2. There is no "quantum isotope effect" between H and D at temperatures in excess of 1500 K achieved in those experiments.

Fig. 6.1 indicates fluid H is a semiconductor from 93 to 140 GPa. From 140 to 180 GPa, electrical conductivity is ~2000/(Ω-cm) (500 μΩ-cm), typical of strong-scattering metals with MMC (Mott, 1972). Completion of the crossover from semiconductor to metal occurs at $P_{met} \approx 140$ GPa. It is not yet possible to measure pairs (ρ, T) achieved in the experiments. At 140 GPa, fluid H has a temperature of 2600 K (Nellis et al., 1999), which is ~three times the melting temperature of H at 140 GPa (Deemyad and Silvera, 2008). Thus atomic correlations in MFH are expected to be negligible well away from melting (Mott, 1934). Fortov et al. (2003) measured similar fluid hydrogen conductivities under QI multiple-shock compression.

Fig. 2.17, calculated with a hydrodynamic simulation code, indicates that more than eight shocks reverberated in liquid H_2 to make MFH at 140 GPa. Model calculations with the ideal-gas EOS shown in Fig. 2.8 indicate that more than eight shocks produce an isentrope, starting from the first shock state on the Hugoniot. Thus, QI multiple-shock compression of liquid H_2 with ρ_0 = 0.071 g/cm^3 was achieved with a sufficient number of multiple shocks to achieve quasi-isentropic compression using sapphire anvils with ρ_0 = 4.00 g/cm^3.

To analyze the semiconductivities to determine ρ_{met} at T_{met}, it is necessary to know ρ and T corresponding to the measured semiconductivities. At present ρ and T generated under multiple-shock compression cannot be measured directly. For this reason ρ and T were calculated with a theoretical EOS, which was constrained by substantial experimental data. The first shock in a sequence of multiple shocks is on the Hugoniot by definition. Experimental data measured under single-shock compression were used as bases with which to develop an appropriate theoretical EOS model to calculate values of ρ and T achieved by multiple-shock compression.

6.3.4 EOS of Dense Fluid Hydrogen Based on Shock-Compression Experiments

Because it is not yet possible to measure (ρ, T) pairs achieved in each shock reverberation experiment in Fig. 2.16, (ρ, T) values were calculated with a theoretical phenomenological EOS model, which was derived based on an extensive body of experimental single-shock data. The theoretical EOS (Holmes et al., 1995; Ross, 1996, 1998) was generated with a measured experimental data base of liquid D_2 and H_2, which included measured single-shock Hugoniot P-ρ data (Nellis et al., 1983), measured single-shock temperatures (Holmes et al., 1995) and shock-reverberation pressures P_f of fluid hydrogen determined by shock-impedance matching the ten conductivity experiments (Nellis et al., 1999).

The model includes an effective pair potential determined by Hugoniot data and is consistent with measured Hugoniot temperatures. For each multiple-shock compression experiment, the effective pair potential was used to calculate the isentrope from the first-shock state up to P_f determined by shock-impedance matching that experiment. First-shock temperature T_H obtained in each experiment is ~40% of final temperatures T_f calculated for each experiment. First-shock temperatures T_H were measured by Holmes et al. (1995) and so uncertainties in calculated values of T_f are comparable to uncertainties in measured values of T_H.

Specifically, for each multiple-shock experiment the initial shock state H_1 in fluid H was obtained by shock-impedance matching measured impact velocity for that experiment. Ross' theoretical EOS was then used to calculate the isentrope of fluid H from H_1 up to P_f for each of the ten conductivity experiments. Calculated values of (ρ_f, T_f) at P_f were taken as peak value of (ρ, T) achieved in each experiment. Measured electrical conductivities σ and associated calculated (ρ, T) of each measured conductivity are published (Nellis et al., 1999). The largest individual contribution to each calculated peak reverberation temperature T_f is $T_1 \approx 40\%$ of T_f, which was caused by the first shock H_1 of each of the ten multiple-shock compressions. Measured experimental data highly constrain those first-shock (ρ_1/T_1) values used to calculate the (ρ_f/T_f) values.

Metallization density ρ_{met} was taken as the density at finite T_{met} at which the semiconducting mobility gap in the electron density of states $E_g(\rho) \rightarrow k_BT$, where k_B is Boltzmann's constant. $E_g(\rho)$ was obtained by fitting measured semiconductivities in the range of 93 to 124 GPa (Nellis et al., 1999) to the standard expression for a thermally activated semiconductor using calculated (ρ,T) values of Ross described previously:

$$\sigma(\rho, T) = \sigma_0 \exp\left[-E_g(\rho)/2k_BT\right], \tag{6.2}$$

where $\sigma(\rho,T)$ is measured electrical conductivity, σ_0 is a constant and $E_g(\rho)$ is the mobility gap in the electronic density of states of fluid H assumed linear in ρ between 90 and 124 GPa, a pressure range away from the nonmetal-metal transition at 140 GPa. The resulting fit for $E_g(\rho)$ is given by

$$E_g(\rho) = 1.22 - 62.6(\rho - 0.30), \tag{6.3}$$

where $E_g(\rho)$ is in eV, ρ is in mol H_2/cm^3 in the range $0.29 < \rho < 0.32$, $\sigma_0 = 90/(\Omega\text{-cm})$ and $E_g(\rho_{met}) \approx k_BT_{met} = k_B$ (2600 K) = 0.22 eV at 0.31 to 0.32 mol $H_2/cm^3 \approx 0.64$ mol H/cm^3 (Weir et al., 1996a; Nellis et al., 1999).

Eq. (6.2) implies that $E_g(\rho)$ determined from the fit is uncertain to the same degree as temperatures T used to generate the fit. To estimate systematic sensitivity of ρ_{met} to calculated (ρ, T) pairs at which conductivities were measured, a second hydrogen EOS generated by Kerley (1983), which was in a form appropriate for use in a hydrodynamic simulation code, was used to calculate the various pairs of ρ, T at which conductivities were measured from hydrodynamic simulation of each experiment. Kerley's EOS in those calculations neglected dissociation in the regime of interest.

Two values of ρ_{met} using both sets of (ρ,T) values calculated with the theoretical models of Ross and of Kerley were determined from the condition $E_{gap}(\rho_{met}) \approx k_B T_{met}$. Although individual values of $E_{gap}(\rho)$ determined with the two EOS models differ systematically by ~20%, the two respective ρ_{met} values differ by only ~2% on extrapolation to the state at which $E_{gap}(\rho_{met}) \approx k_B T_{met}$, sufficiently accurate to determine ρ_{met}. Results obtained using pairs of (ρ,T) values calculated with the theoretical model of Ross are illustrated in Fig 6.2. Preferred values of densities and temperatures in the conductivity experiments were calculated with Ross' model because they were highly constrained by experimental shock-compression data, which was not true of the other EOS used. Uncertainties in calculated pressures, densities and temperatures in those multiple-shock compression experiments are estimated to be 2%, 5% and 10%, respectively.

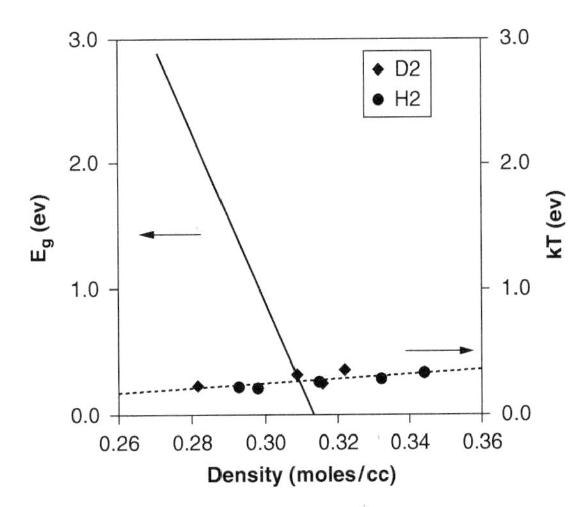

Fig. 6.2. Electronic mobility gap $E_g(\rho)$ of fluid hydrogen/deuterium obtained from least-squares fit of measured electrical semiconductivities from Fig. 6.1 plotted versus associated calculated density of hydrogen/deuterium in units of diatomic moles/cc. Dotted line is plot of calculated $k_B T$ versus ρ. Pairs of (ρ,T) at associated conductivities were calculated with Ross' hydrogen EOS described in text (Weir et al., 1996a; Nellis et al., 1999). In region in which $E_g(\rho) \approx k_B T$, hydrogen completes crossover or continuous transition from semiconducting to metallic fluid. Copyright 1996 by American Physical Society.

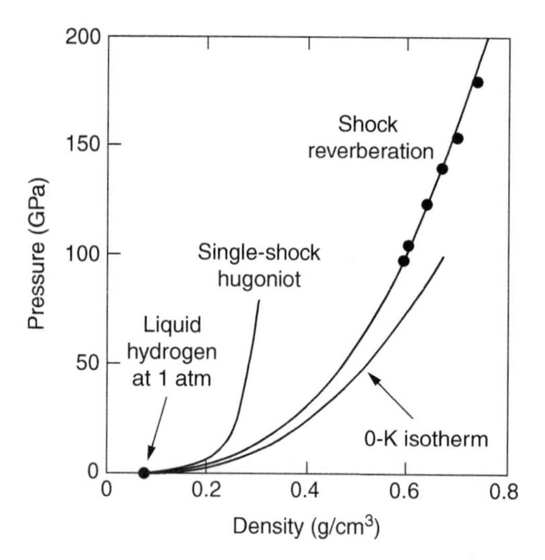

Fig. 6.3. Comparative plots of pressure-density curves achieved with shock reverberation, shock compression (Hugoniot) and 0-K isotherm. Black dots are states in Fig. 6.1 (Nellis et al., 1999). Copyright 1999 by American Physical Society.

Fig. 6.3 illustrates various types of compression: (1) QI compression of fluid H by shock reverberation, (2) adiabatic shock compression on the Hugoniot and (3) the 0-K isotherm of solid H_2, essentially what is achieved in a DAC. Temperatures and densities on the Hugoniot are too high and low, respectively, to make a metal. Shock reverberation history achieves appropriate amounts of pressure, density, temperature and entropy to make MFH.

6.3.5 Nature of Metallic Fluid Hydrogen

Because P_{met} (140 GPa) and T_{met} (2600 K) differ substantially from initial P_0 (10^{-4} GPa) and T_0 (20 K), initial liquid H_2 molecules transform into something other than diatomic fluid H_2 at extreme P and T. Atomistic computational simulations at (ρ_{met},T_{met}) have been calculated on a fs timescale; bulk electrical conductivities have been measured on a ns timescale. The preponderance of those results indicate that MFH is probably monatomic near metallization density ρ_{met} = 0.64 mol H/cm^3.

Ab initio molecular dynamics calculations (AIMD) and Kubo-Greenwood conductivity calculations have shown that at metallization density ρ_{met} (r_s = 1.5) and temperature T_{met} (3000 K), monatomic fluid H is clearly a metal. Transient dimers with lifetimes of ~10^{-14} s are observed in those calculations, which lifetime is comparable to a single vibrational period of a free H_2 molecule. Those brief lifetimes of close approach of pairs of H atoms (~10^{-14} s) are thermal fluctuations at high densities, rather than the formation of an H_2 molecule (Pfaffenzeller and Hohl, 1997).

The high-pressure metallic fluid phase of fluid hydrogen has been suggested to occur possibly by band overlap of short-lived transient H_2 dimers (Nellis et al., 1998), which appears a remote possibility based on calculations cited previously, although dense fluid hydrogen is a complicated system and the possibility exists.

Conductivities have been calculated with coupled electron-ion Monte Carlo simulations. Calculated dc conductivity at 3000 K and several densities shows a fluid semiconductor-to-metal transition at densities in the range of $1.60 < r_s < 1.65$ and temperature of 3000 K (Lin et al., 2009), in good agreement with electrical conductivities measured under dynamic compression. The molecular-atomic transition curve of liquid hydrogen has been mapped at high pressures P and temperatures T. The relatively narrow phase curve passes through the point (P,T) = (100 GPa/2000 K), above which the liquid is atomic and below which the liquid is diatomic (Tamblyn and Bonev, 2010a, 2010b). The observed value of electrical conductivity of MFH is comparable to the value calculated theoretically at 0.6 g/cm^3 and 3000 K with ab initio molecular dynamics (Holst et al., 2011). Subramanian et al. (2011) have measured vibron spectra of H_2 in DAC at pressures of ~120 GPa and temperatures up to 950 K. The relatively narrow H_2 vibron line at 300 K broadens substantially above 800 K. The results are ascribed to modification of intramolecular bonding and might be a precursor to molecular dissociation at higher temperatures.

Systematics of measured electrical conductivities of fluid H, N and O under QI compression (Bastea et al., 2001; Chau et al., 2003a) and of fluid Rb and Cs heated at static pressures (Hensel and Edwards, 1996; Hensel et al., 1998; Edwards et al., 1998) are plotted in Fig. 6.4, as logarithm of measured conductivity versus $D^{1/3}a^*$, where D is atomic density and a* is effective Bohr radius of atomic wave functions. Fluid Cs and Rb in those experiments were heated well above their melting temperatures to 2000 K. On melting, solid metallic Cs and Rb become semiconductors and must be compressed to become metallic fluids. Temperatures of metallic fluids in Fig. 6.4 were 2000 K for Rb and Cs, 1700 K to 3000 K for H and more than 5000 K for N and O. At such high temperatures it is quite reasonable to expect complete dissociation of those diatomic liquids.

Fig. 6.4 shows that as adjacent atoms come closer together under pressure, relative to their respective Bohr radii, conductivity increases by increased overlap of electron wave functions on adjacent atoms until maximum electron carrier density is reached. Electron scattering is strong in a fluid and so electrical conductivity reaches a maximum of MMC (Mott, 1972). MMC corresponds to the state for which electron mean-free path is comparable to inter-atomic distance, and thus

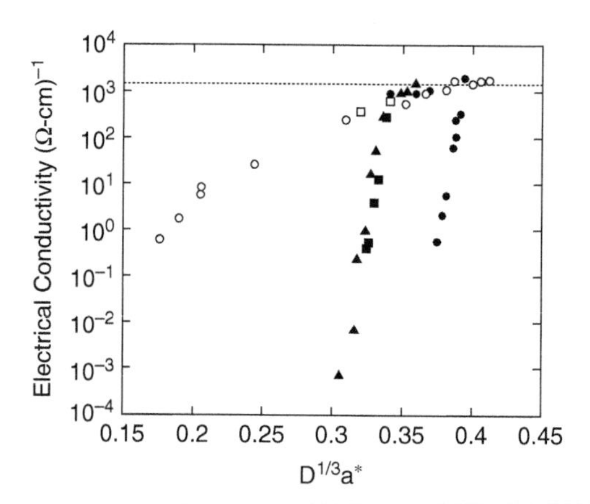

Fig. 6.4. Logarithm of electrical conductivity versus $D^{1/3}$a*, where D is atomic density and a* is effective Bohr radius of outermost atomic wave functions: open circles and squares: Cs and Rb, filled diamonds and squares: O and N, filled circles: H (Chau et al., 2003a). Copyright 2003 by American Physics Society.

MMC depends weakly on element and temperature. Electrical conductivities with scaled densities of monatomic H, Rb and Cs show these elemental fluids become metallic at $D^{1/3}$a* $\approx$ 0.38. The polyvalent low-Z fluids N and O become metallic at $D^{1/3}$a* $\approx$ 0.34.

The rate of increase of electrical conductivity of respective adjacent atoms depends on the radial extents of wave functions of outer electrons of the respective elements. For this reason, Fig. 6.5 illustrates effects of screening by inner-shell electrons on radial extents of conducting outer electrons. Calculated radial extents of wave functions of the five elements (Chau et al., 2003a) are illustrated in Fig. 6.5. The outer electrons of H, Rb and Cs are spherically symmetric s^1 electrons. Wave functions of the outer electrons of N and O were angle-averaged (Herman and Skillman, 1963). The $1s^1$ wave function of the single electron of the H atom is unscreened and the most tightly bound in Fig. 6.4. The outermost three and four electrons of the N and O atoms, respectively, are screened by the their inner $1s^2$ [He] cores. The $5s^1$ outermost electron of the Rb atoms is screened by its [Kr] core. The $6s^1$ outermost electron of the Cs atom is screened by its [Xe] core. The simple systematics of the elemental fluids illustrated in Figs. 6.4 and 6.5 suggest fluid H, N, O, Rb and Cs at temperatures of 2000 K and higher are monatomic metals.

Logarithms of electrical conductivities versus shock-reverberation pressure P_f have also been measured for Li (Bastea and Bastea, 2002). Li is a solid metal with high electrical conductivity at ambient, which decreases at high dynamic pressures in the fluid under shock reverberation. Above 100 GPa, conductivities of fluid Li

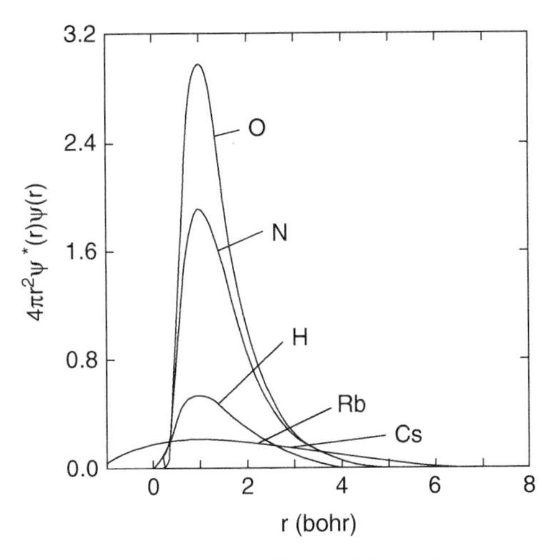

Fig. 6.5. Radial charge density distributions of five elements in Fig. 6.4 plotted versus radius (in Bohr). Peaks of all curves have been translated to common radius to illustrate radial extents of various curves. Note how radial extents of wave functions of H, of N and O and of Rb and Cs systematically increase radially outward with increasing atomic number Z. Calculated charge densities of Rb and Cs appear to extend to negative radii only because those curves have been shifted as described (Chau et al., 2003a). Copyright 2003 by American Physical Society.

are within a factor of two to three of the conductivities of metallic fluid H, N and O. Under multiple-shock compression above ~100 GPa, all four fluid metals reach MMC, independent of whether MMC is approached from an initial state that is electrically insulating or metallic (Nellis, 2006a).

6.4 Metallic Fluid H in a Diamond Anvil Cell

Pressures from 100 to 170 GPa at temperatures from 1000 to 2500 K in fluid hydrogen have recently been obtained in DACs. These conditions are comparable to *P*/*T* at which electrical conductivities of fluid H were measured under dynamic compression (Nellis et al., 1999) and plotted in Fig. 6.1. The question of a liquid-liquid phase transition (LLT) in dense hydrogen is an issue of long-standing importance in theoretical physics (Norman et al., 2015, and references therein). This LLT is often known as a plasma phase transition (PPT), which is a way to investigate MFH that is complementary to dynamic compression that induces dissociation and metallization by minimizing free energy of the fluid.

In 2010 density functional theory and quantum Monte Carlo calculations at high *P*/*T* showed evidence for a first-order phase transition in fluid hydrogen between a molecular phase with low electrical conductivity and an atomic phase with high electrical conductivity. The critical point of this transition was estimated to be in the vicinity of 2000 K and 120 GPa (Morales et al., 2010; Tamblyn and Bonev, 2010a, 2010b; Lorenzen et al., 2010). The phase line along which fluid hydrogen reaches the fully dissociated electrically conducting monatomic phase provides an opportunity to investigate MFH. At temperatures above the critical point, a continuous dissociative crossover is expected from a low-conductivity molecular phase to the high-conductivity atomic phase, while below the critical point in the two-phase region between those two phases a discontinuous first-order transition is predicted.

Studying fluid H at high *P*/*T* in a DAC is challenging because hydrogen diffuses rapidly out of a DAC at high *P*/*T*. To inhibit hydrogen diffusion out of a DAC, sample heating was achieved with laser pulses ~280 ns long applied at a frequency of 20 kHz and other measures. Static high pressures in fluid hydrogen were obtained thusly in the range 100 to 170 GPa at pulsed temperatures up to 2200 K (Dzyabura et al., 2013; Zaghoo et al., 2016).

Optical transmittance and reflectance were measured, which suggest a first-order phase transition accompanied by changes in transmittance and reflectance characteristic of a metal. At 170 GPa and 1250 K, electrical conductivity of fluid H derived from measured optical properties is $(2.1\pm1.3)\times10^3/(\Omega\text{-cm})$ in agreement with Mott's MMC measured under dynamic compression above 140 GPa (Fig. 6.1). At highest temperatures and metallic H film thicknesses, reflectance saturates at 0.55, consistent with optical reflectance at high temperatures measured from shock fronts in shock-compressed fluid deuterium generated with a high-intensity pulsed laser (Celliers et al., 2000). The observed phase line along which optical transmittance and reflectance were measured has a negative *T*-*P* slope in agreement with theories of the PPT. *P*/*T* points measured along the PPT phase line are plotted in Fig. 6.6 (Dzyabura et al., 2013; Zaghoo et al., 2016).

Ohta et al. (2015) performed similar hydrogen experiments in a DAC at pressures in the range 82 to 106 GPa and temperatures in the range 2200 to 2500 K. Those data are also plotted in Fig. 6.6. As seen in this figure, the results of Ohta et al. are at somewhat higher temperatures than those of Zaghoo et al. but essentially along the same phase line. Both data sets lie on a common curve characteristic of the PPT.

Also plotted in Fig. 6.6 are *P*/*T* points at which electrical conductivities of fluid H samples were measured under dynamic compression, along with values of electrical conductivities measured at those *P*/*T* points (Nellis et al., 1999). Agreement between the experiments and theory is remarkable. In Fig. 6.6, electrical conductivities measured under dynamic compression near ~1700 K and ~100 GPa

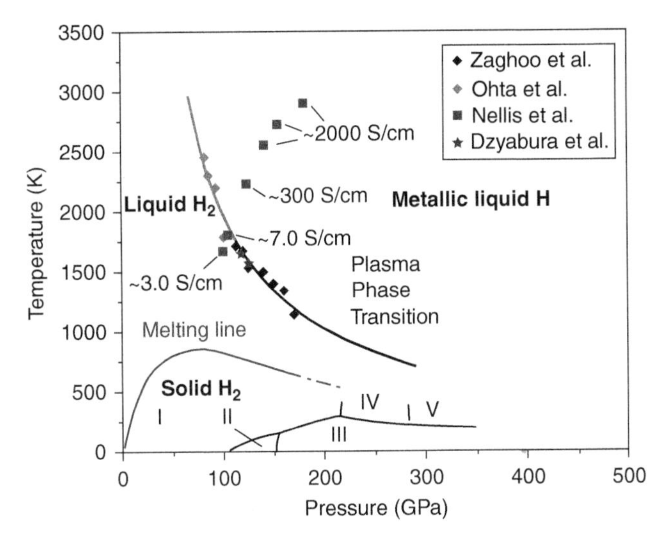

Fig. 6.6. Equation of state of dense hydrogen showing *P*/*T*s (squares) at which electrical conductivities of fluid hydrogen were measured under dynamic compression (Nellis et al., 1999; Fig. 6.1). Data of Fortov et al., 2003 are similar. Also shown is phase line of PPT measured in DACs (Dzyabura et al., 2013; Ohta et al., 2015; Zaghoo et al., 2016). Melting line from Deemyad and Silvera (2008). Electrical conductivities measured under dynamic compression at various squares are indicated. Under dynamic compression semiconductivities increase exponentially from 90 to 140 GPa. Conductivities are ~constant at ~2000 S/cm (Mott's MMC) from 140 to 180 GPa. Copyright 2016 by American Physics Society.

are ~3 S/cm and 7 S/cm, which is low as expected for liquid H_2 below the PPT phase line measured by Dzyabura et al. (2013), Ohta et al. (2015) and Zaghoo et al. (2016). Under dynamic compression of fluid H above the measured PPT phase line measured electrical conductivities increase rapidly from ~10 S/cm up to 2000 S/ cm at which they plateau at MMC of a poor metal. Measured electrical conductivities of fluid H under dynamic compression of Fortov et al. (2003) are in good agreement with those of Nellis et al.

A question has been raised as to whether metallization of fluid hydrogen occurs via a first-order phase transition or a continuous crossover (McWilliams et al., 2016). Fig. 6.6 indicates that both occur – that is, there are two transitions by different mechanisms. Under fast QI multiple-shock compression, fluid H undergoes a continuous transition from a semiconductor to a poor metal with MMC. Electrical conductivities are measured directly under dynamic compression. States achieved are determined by the hydrodynamics of shock reverberation and minimization of free energy. Metallization of H occurs at the density predicted by WH (1935). In a DAC, static pressure at a given point is essentially fixed and a first-order transition is observed from optical properties as the phase line is crossed by varying the intensity of incident optical radiation.

6.5 Metallic Solid H in a Diamond Anvil Cell

Metallic solid hydrogen (MSH), has recently been reported in a sample at a static pressure of 495 GPa and a temperature of 5.5 K in a diamond anvil cell (Dias and Silvera, 2017). Metallization was detected by measurements of optical reflectance. This is the first reported observation of metallic solid H predicted by Wigner and Huntington in 1935. This remarkable experimental observation needs to be reproduced. Pressure and temperature differ dramatically between 140 GPa/3000 K and ~500 GPa/5.5 K for MFH and MSH, respectively. Metallic solid H has been predicted to be a high-temperature superconductor (Ashcroft, 1968).

Although an absolute static pressure scale above ~200 GPa has yet to be developed, the pressure range above was estimated by linear extrapolation of the diamond-Raman pressure scale (Akahama and Kawamura, 2007). A calibrated pressure scale is needed up to 500 GPa down to 5K. The estimated electron density of MSH of $7.8\times10^{23}/cm^3$ is consistent with H atom density of $4.0\times10^{23}/cm^3$ in degenerate metallic fluid H at about half the density of MSH.

The possibility remains that MSH might actually be a fluid. In addition to the existence of degenerate MFH at higher temperatures (Nellis et al, 1999), metallic H at low temperatures has been predicted theoretically to be a liquid metal and a liquid superconductor (Jaffe and Ashcroft, 1981; Babaev et al., 2004; Bonev et al., 2004). Whether MSH at ~500 GPa is a solid or a fluid is an interesting question in quantum physics and a question that needs to be answered experimentally.

6.6 Dynamic Compression of Hydrogen: Z Accelerator

Liquid D_2 has been compressed quasi-isentropically in two stages by current-generated magnetic pressure at the Z accelerator (Knudson et al., 2015). The first stage generated three shocks by reverberation in liquid D_2. The second stage then drove a ramp wave that further compressed deuterium isentropically up to ~300 GPa at calculated peak temperatures in the range 800 to 1700 K. Diagnostics were optical interferometric VISAR signals to measure velocity histories $v(t)$, which were driven by pressure history $P(t)$ in deuterium. $P(t)$ was derived from $v(t)$ with a hydrodynamic computer code. Density and temperature histories, $\rho(t)$ and $T(t)$, were calculated theoretically. Calculated $T(P)$s were nearly constant between 150 and 300 GPa, an unusual result that warrants experimental verification by temperature measurements. Reflectivities were measured up to ~300 GPa at which reflectivity of fluid D increases abruptly to ~45%, which suggests an electrical conductivity of a few 10^5 S/m, typical of MMC.

H densities are greater than 1.0 mol H/cm^3 at 300 GPa, factors of ~2 and ~1.5 greater than P and ρ respectively, on completion of the crossover at 140 GPa at the 2SG. These substantial differences suggest these transitions in the Z and 2SG experiments are different transitions. These Z results are interesting and the nature of the transition at 300 GPa needs to be clarified.

7

Unusual Magnetic Fields of Uranus and Neptune: Metallic Fluid H

In the 1980s, NASA's Voyager 2 spacecraft observed the unusual non-dipolar non-axisymmetric magnetic fields of Uranus and Neptune (U/N), which are unique in the solar system. Planetary magnetic fields are made by dynamos: convection of electrically conducting fluids across lines of magnetic flux (Stevenson, 1983). *P* and *T* in U/N range from ~70 K at their surfaces up to ~700 GPa and ~7000 K near their centers (Hubbard, private communication, 2011). The explanation of those unusual magnetic fields has been a major unanswered question in planetary science ever since their discovery. The likely explanation must be consistent with what is known experimentally on Earth about likely planetary fluids at likely *P* and *T* in U/N at which those fields are probably made, as well as with radial density distributions of U/N derived from gravitational data measured by Voyager 2 (Helled et al., 2011). Over the past three decades an enormous database has been measured on Earth of likely planetary fluids at likely *P* and *T* in U/N (Nellis, 2015b). This chapter is about using that dynamic compression data to develop a likely explanation of those puzzling planetary magnetic-field observations.

In 1977, NASA's Voyager 2 spacecraft departed Earth to visit the Giant Gas planets Jupiter and Saturn and then on for flybys of the Giant Ice planets Uranus and Neptune (U/N). It took Voyager 2 nine years to arrive at Uranus and another three years to arrive at Neptune in 1989. Voyager 2 measured the magnetic and gravitational fields of U/N on the respective day it flew by each of them. As of this writing in 2016, Voyager 2 continues to measure magnetic fields in interstellar space and transmits that information back to Earth as it travels outward at ~16 km/s relative to the sun (U.S. National Aeronautics and Space Agency Voyager Program website).

A velocity of 16 km/s is twice maximum velocity of an impactor launched by the 2SG at LLNL and comparable to velocities achieved with the Z Accelerator at SNLA. Dynamic compression by impact at those velocities achieves *P/T*s in fluids that are expected in planetary interiors. Impact velocities up to 8 km/s achieve *P/T*s that are representative of depths in U/N at pressures and temperatures up to ~300 GPa and few 1000 K, conditions at which external magnetic fields are generated that could be measured by Voyager 2. For this reason, material properties measured on Earth are relevant to understanding magnetic-field generation in U/N (Nellis, 2015b, 2017), in Jupiter (Nellis et al., 1995, 1996) and in exo-planets.

The fact that magnetic fields of U/N are not nearly dipolar and not nearly axially symmetric, as is Earth's magnetic field, has generated substantial computational modeling to try to explain the generation of those fields (Ruzmaikan and Starchenko, 1991; Podolak et al., 1991; Starchenko, 1993; Hubbard et al., 1995; Holme, 1997; Stanley and Bloxham, 2004, 2006; Helled et al., 2011). The general approach has been to construct magneto-hydrodynamics (MHD) models that compute magnetic field shapes similar to those observed for U/N by assuming a specific geometry of planetary interiors that computes magnetic field shapes similar to those observed for U/N. Although the material model chosen for the interiors of U/N is not inconsistent with radial density distributions derived from measured gravitational harmonics, the gravitational data measured by Voyager 2 are unable to support the specific computational requirement that the unusual magnetic fields of U/N are made in a convective spherical annulus whose thickness is small compared to radii of U/N. That assumption is non-unique with respect to possible causes of those unusual fields. Such detailed assumptions about interior planetary structures and conductivities used in those MHD calculations cannot be verified to be true.

On the other hand, an extensive database of measured material properties of possible fluids in the Ice Giants has been measured with dynamic compression at representative interior planetary pressures and temperatures. In this work, a likely picture of the interiors of U/N is developed that is consistent with the natures of observed magnetic field shapes and with properties of likely planetary fluids measured at extreme conditions. In developing this picture, consideration of deep-Earth seismic analyses by Hide et al. (2013) suggests the mechanism by which rotational motion of Earth's rigid, rock mantle is coupled into convective motions of Earth's fluid-Fe outer core in which the magnetic field of Earth is generated. Because there is no apparent corresponding mechanism that strongly couples rotational motion of U/N into convective motions that generate magnetic fields of U/N, the fields of U/N are probably relatively free to wander as fluctuations in local convective fluid flows and temperatures dictate, essentially unconstrained by rotation of U/N. This analysis suggests statistical distributions of convective flow patterns, compositions and electrical conductivities would be a more representative way with which to model dynamos of U/N.

7.1 Chemical Compositions and Properties of Uranus and Neptune

U/N have similar sizes, densities, temperatures, magnetic and gravitational fields and probably chemical compositions, although precise internal compositions are unknown. U/N are probably composed of (1) Gas, an H-rich mixture of H and He, (ii) Ice, a mixture primarily of H_2O, CH_4 and NH_3 and (3) Rock, a mix of silicates and Fe/Ni (Stevenson, 1982; Hubbard, 1984; Irwin, 2003). The envelops of U/N are primarily H and He with solar abundances of about 92 at.% and 7 at.%, respectively (Arnett, 1996). The remaining 1 at.% is other elements. In space, "icy" molecules are called "nebular" – very low-density molecular gases. Planetary "ices" are hydrogenous fluids with relatively abundant reactive elements O, N or C. Planetary ice refers to chemical composition, rather than to a solid state of matter. Phase diagrams of H_2O and NH_3 have been calculated at pressures and temperatures in the ranges 30 to 300 GPa and 300 to 7000 K, respectively (Cavazonni et al., 1999).

U/N have masses 14.5 and 17.2 M_E, respectively, where M_E is mass of Earth. Total mass of the giant planets Jupiter, Saturn, Uranus and Neptune is 445 M_E. Total mass of the rocky planets Earth, Venus, Mars and Mercury is 2.0 M_E. The most abundant mass of the solar system is MFH in Jupiter and Saturn. U/N are very H-rich because of their gas envelops and decomposition of ices at high interior *P/T*. Dynamic experiments on planetary fluids began in earnest when Voyager 2 departed Saturn for U/N (Hubbard, 1981; Mitchell and Nellis, 1982).

Accretion of massive amounts of hydrogen and ices generates high gravitational pressures *P* and associated temperatures *T*. Interior *P* and *T* in U/N range up to several 100 GPa (10^6 bar = Mbar) and several thousand K. Pressures up to ~200 GPa under QI compression are achieved in thermally equilibrated planetary fluids by dynamic compression on Earth, as discussed for hydrogen in Chapter 6. Ideally, one would like to know which fluids at which *P*/*T* to study. However, the answer is not known a priori to either question.

One way to generate a reasonable qualitative model for the interiors of U/N is to collect on Earth a database of measured properties of representative planetary fluids at representative expected internal *P/T*. Measurements that have been made at high *P/T* include electrical conductivities, equations of state, thermal emission temperatures and optical scattering experiments of dense fluid hydrogen, He, ices, and so forth. Such experimental information splits naturally into two subgroups: (1) H_2 and He, which comprise the envelopes, and (2) ices, along with silicates, Fe and Ni, which comprise the cores. Qualitative conclusions herein about U/N based on the substantial database measured on Earth do not depend sensitively on exactly which fluids at exactly which *P/T* exist in their cores.

The boundary regions in U/N between H/He envelopes and ice cores occur around ~100 GPa and ~2000 K, above which shock compression experiments indicate small-molecular fluids dissociate and, thus, probably re-react to form new chemical species other than compressed and heated accreted nebular molecules. That is, there probably is no planetary ice in the Ice Giants. The distinctly different chemical compositions of U/N relative to those of Earth and of Jupiter/Saturn play a major role in generating the unusual magnetic fields of U/N.

7.2 Voyager 2's Uranus and Neptune

Voyager 2 measured gravitational and magnetic fields of U/N as it flew by both planets. Those first measurements near U/N in space, together with measurements on Earth of representative fluids, are extremely valuable in deriving likely pictures of the interiors of the Ice Giants. Measured external magnetic fields of U/N are ~2×10^{-5} Tesla (Stevenson, 2010), which is comparable to Earth's field. If the measured fields of U/N are force fit to effective dipolar fields, the effective magnetic axes are tilted 59° and 47° from their respective rotational axes and the effective dipole centers are offset by ~0.33 R_U and ~0.55 R_N, respectively, from the physical centers of U/N, where R_U and R_N are the radii of U and N, as illustrated in Fig. 7.1.

Those external magnetic fields are probably made primarily by degenerate, MFH (Stevenson, 1998; Nellis, 2015b). In comparison the magnetic fields of Earth, Saturn and Jupiter are essentially dipolar with their magnetic axes aligned within ~10 degrees of their respective rotational axes (Stevenson, 2010). Thus, a possible clue to the explanation of the magnetic fields of U/N might lie in the fact that whatever causes approximate alignment of magnetic field and rotational axes of Earth does not occur in U/N. Or alternatively an important question is whether

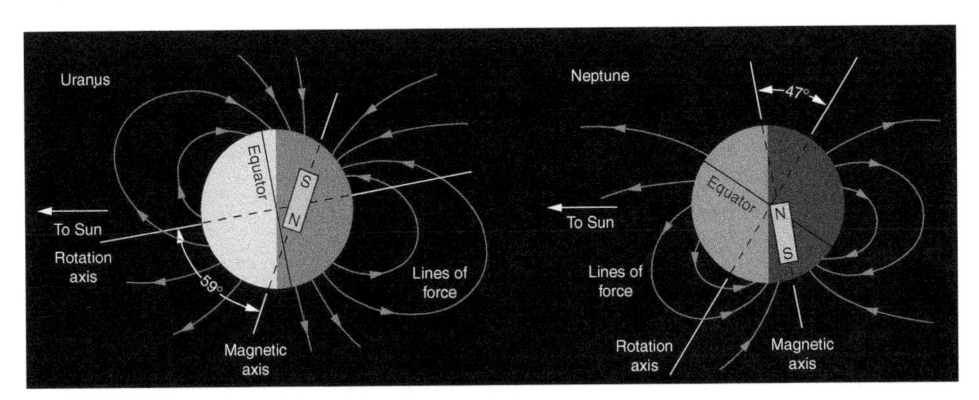

Fig. 7.1. Non-dipolar non-axisymmetric magnetic fields of Uranus and Neptune measured by Voyager 2 spacecraft of U.S. National Aeronautics and Space Agency in 1986 and 1989, respectively. Copyright © U.S. NASA.

there are possible experimentally observed bulk phenomena on Earth that might reasonably be expected to be relevant to aspects of the observed magnetic fields of the Ice Giants?

7.3 Dynamic Compression Experiments on Planetary Fluids

The vast majority of measured properties of fluids at high planetary *P*/*T*s have been made under dynamic compression by adiabatic single-shock compression and by QI multiple-shock compression. These experiments are discussed in Chapters 2 and 6 and Nellis (2006a). Reviewing briefly, the front of a shock wave travels at supersonic velocity in the medium ahead of the front. Because a shock front is supersonic, material ahead of a shock front receives no signal of impending compression until the shock front itself arrives and "snowplows" that material. In this way, shock propagation in a fluid compresses rapidly and adiabatically, which generates dissipation T and entropy S in a thin (~ps and ~nm) shock front (Hoover, 1979). The state achieved in planetary fluids by dynamic compression is thermally equilibrated.

Higher density ρ and lower temperature T are achieved relative to that of a single shock by choice of applied dynamic pressure pulse. Specifically, QI compression is achieved by tailoring rise time of the wave front to be longer than that of a sharp shock. Common techniques to this end are multiple-shock and ramp-wave compression. By these means thermodynamics are tunable over a wide range by tuning the hydrodynamics of compression.

7.3.1 Experimental Data for H_2, He and Ices and Radial Density Distributions of U/N

Liquids investigated under multiple-shock compression include H_2/D_2 (Weir et al., 1996a; Nellis et al., 1998; Nellis et al., 1999; Fortov et al., 2003; Nellis, 2013), He (Fortov et al., 2003), O_2 (Bastea et al., 2001; Chau et al., 2003a), N_2 (Chau et al., 2003b), H_2O (Yakushev et al., 2000; Chau et al., 2001), S (Yakushev et al., 2000; Mintsev and Fortov, 2015) and synthetic Uranus (SU) (Chau et al., 2011). SU is a single-phase liquid mixture of polar molecules H_2O, NH_3 and C_3H_8O (isopropanol).

Liquids investigated under single-shock compression include He (Nellis et al., 1984a), H_2/D_2 (Nellis et al., 1983; Nellis et al., 1992; Holmes et al., 1995; Holmes et al., 1998; Nellis, 2002b; Boriskov et al., 2005), N_2 (Nellis and Mitchell, 1980; Nellis et al., 1984b; Radousky et al., 1986; Nellis et al., 1991a); O_2 (Nellis and Mitchell, 1980; Hamilton et al., 1988a), CO (Nellis et al., 1981), CO_2 and air (Nellis et al., 1991b), CH_4 (Nellis et al., 1980; Nellis et al., 1981; Radousky et al., 1990; Nellis et al., 2001), NH_3 (Mitchell and Nellis, 1982; Radousky et al., 1990),

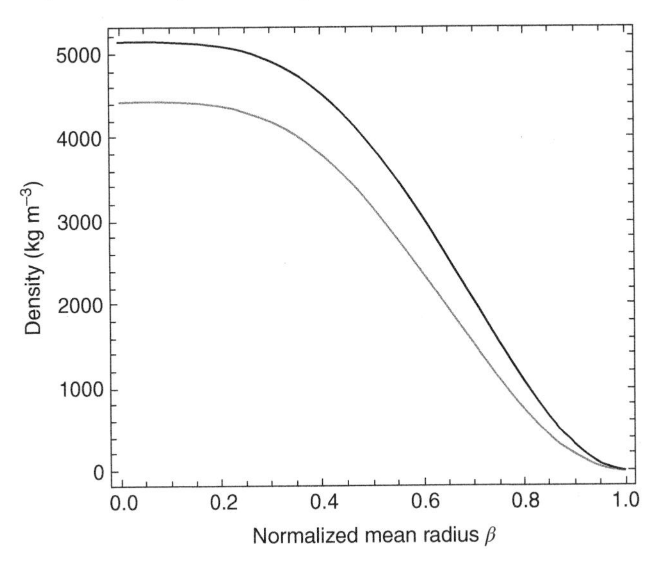

Fig. 7.2. Calculated radial densities normalized to mean radii of Uranus (gray) and Neptune (black). (Helled et al., 2011) © 2011 American Astronomical Society. Reproduced with permission.

H_2O (Mitchell and Nellis, 1982; Lyzenga et al., 1982; Holmes et al., 1985), C_6H_6 (Nellis et al., 1984c; Nellis et al., 2001), C_4H_8 (Hamilton et al., 1988b; Nellis et al., 2001), CH_2 (Nellis et al., 1984c) and SU (Nellis et al., 1997). Hydrocarbons in general decompose at high pressures and temperatures generated by shock compression (Ree, 1979).

That body of experimental results indicates that, above dynamic pressures and temperatures of ~50 to 100 GPa and ~2000 K, respectively, all those small molecules decompose. Pure diatomic molecules become monatomic and then metallic above ~100 GPa. Polyatomic molecules decompose into ions, atoms and perhaps smaller molecules, which probably re-react to form new chemical species. Those experimental results imply that little to no nebular molecules, other than H_2 in their outer envelops, exist in the Ice Giants. Systematics in those laboratory results is a basis for speculation on the general natures of interiors of U/N.

The reason for this "convenient" situation is based on the previous laboratory data plus the radial mass-density distributions derived from measured gravitational harmonics of U/N determined by Voyager 2, which are plotted in Fig. 7.2. The corresponding calculated radial pressure distributions are plotted in Fig. 7.3 (Helled et al., 2011). Fig. 7.2 suggests low-density H-He envelopes and high-density icy cores are at radii greater than and less than as much as ~0.9 R_U and R_N, respectively.

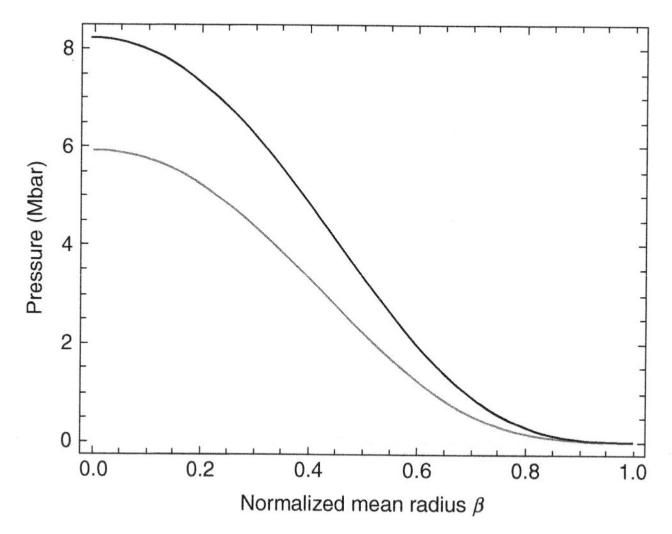

Fig. 7.3. Calculated radial pressures normalized to mean radii of Uranus (gray) and Neptune (black) (Helled et al., 2011). Note that pressures at normalized mean radii of $\beta = 0.9$ are zero, which is unrealistically small given that corresponding densities at same radii in Fig. 7.2 are ~0.6 g/cm^3. © 2011 American Astronomical Society. Reproduced with permission.

While the radial density distributions determined by Voyager 2 are reasonably accurate, EOSs used historically to obtain pressures derived from densities determined by Voyager 2 have unavoidable substantial systematic uncertainties. The EOSs that provide estimates of pressures in U/N are based on assumed materials and theoretical calculations. However, neither chemical compositions nor thermodynamic states and associated chemistry of those materials at high *P/T* are known.

The pressure dependence of measured electrical resistivities of fluid H/D are plotted in Fig. 6.1. Similar conductivities of liquid H_2, dense gaseous H_2 and He have been measured under multiple-shock compression achieved with shock waves generated with chemical explosives (Fortov et al., 2003). The advantage of knowing the pressure dependence of electrical conductivities of fluid H is that known pressure dependences of density can be used to estimate radial positions in U/N at which their external magnetic fields are likely made. Depth in a planet at which magnetic field is made has a significant effect on the shape and spatial distribution of planetary magnetic fields. For example, magnetic fields made close to outer planetary surfaces are likely to have significant quadrupolar and higher contributions.

The reason for this statement is that external magnetic fields of spherical planets can be expanded in terms of spherical harmonics (Graznow, 1983). In this case the magnetic field can be decomposed into dipolar, quadrupolar, octopolar, etc. terms

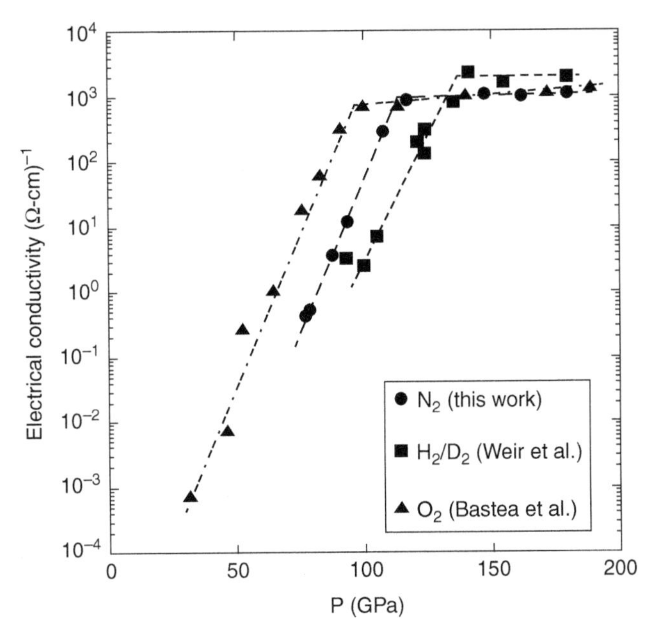

Fig. 7.4. Measured electrical conductivities achieved by QI multiple-shock compression of fluid H/D, N and O versus peak pressure P_f achieved in each (Chau et al., 2003a). Copyright 2003 by American Physical Society.

with corresponding multipolar-field contributions proportional to r^{-3}, r^{-4}, r^{-5}, etc. Thus, relatively larger magnetic field contributions are probably made relatively close to outer planetary radii.

Fig. 6.1 illustrates that dense fluid H undergoes a crossover from semiconductor to metal with MMC at pressures from 90 to 140 GPa; corresponding calculated temperatures range from ~1700 to 2600 K. Measured electrical conductivities increase from 1 to 2000/(Ω-cm) in this range of pressure and temperature. From 140 to 180 GPa measured conductivities of dense H are ~2000/(Ω-cm) (Weir et al., 1996a; Nellis et al., 1999). Historically, electrical conductivities of Ices of ~20/(Ω-cm) have been considered large (Nellis et al., 1988a).

Measured electrical conductivities of liquid H_2/D_2, N_2 and O_2 under QI multiple-shock compression are plotted in Fig. 7.4 (Chau et al., 2003a). These experimental results indicate that these fluids are semiconducting below ~100 GPa and a few 1000 K. Pure H can make contributions to magnetic fields of U/N above ~90 GPa, which is expected in the lower envelopes of U/N. The main point is that non-dipolar magnetic fields of U/N are probably made much closer to their outer surfaces than thought previously (Nellis et al., 1988a), which is very relevant to the unusual magnetic field shapes of U/N. Conductivity curves of fluid H, N and O are similar because of similar electronic structures and strong disorder scattering of

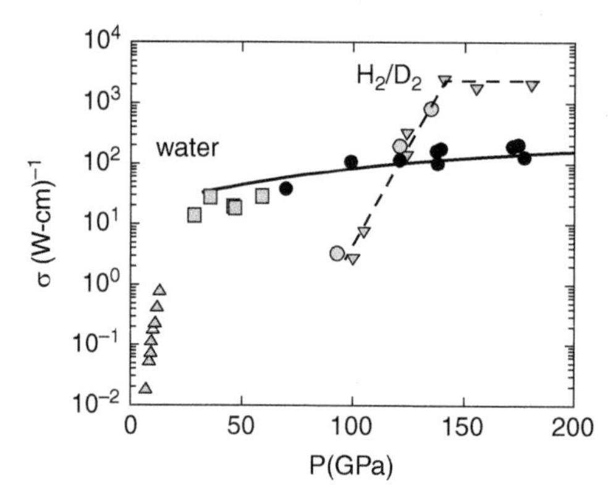

Fig. 7.5. Measured electrical conductivities achieved by QI multiple-shock compression of fluid H/D and water (Chau et al., 2001). Copyright 2001 by American Institute of Physics.

electrons in fluids. The similar conductivities of fluid H, N and O in Fig. 7.4 imply these monatomic fluid metals are probably mutually soluble in the cores to the extent elemental fluids are present in the cores.

Electrical conductivities of fluid H/D and water are plotted in Fig. 7.5. Conductivity of SU (not shown) is within a factor of two of that of H_2O (Chau et al., 2011). Chemical ionization of H_2O probably causes positive protons to dominate conductivity below ~30 GPa (Hamann and Linton, 1966; Mitchell and Nellis, 1982; Yakushev et al., 2000). At higher pressures, it appears electron carriers delocalize from OH^- ions with increasing pressure, which causes electrical conductivity to increase slowly with increasing pressure. Fig. 7.5 suggests that above a pressure of ~300 GPa, conductivities of MFH and water might be comparable. The inference for U/N is that once accreted, nebular molecules sink to depths in the cores at which they decompose and re-react to form other chemical species. At depths in U/N at pressures greater than ~300 GPa most species might be monatomic because densities and temperatures might be too large for chemical bonds to exist at such high compressions and temperatures.

7.3.2 Hydrogen and Water under Static Compression and Heating

The melting curve of water ice has been measured in the range 20–90 GPa with melt temperatures between 1000 and 2400 K (Schwager et al., 2004). Chemically ionized H_2O is observed up to 56 GPa and 1500 K in a DAC (Goncharov et al., 2005). Using a pulsed-heating technique Deemyad and Silvera (2008) measured melting temperatures of hydrogen in the pressure range 50 to 80 GPa with a maximum of 1050 K at 65 GPa. Dzyabura et al. (2013) achieved pressures and temperatures in fluid hydrogen in the ranges 1000 to 1700 K at 119 GPa and in the range 900 to 1500 K at 125 GPa. The plasma phase transition in dense fluid hydrogen has been reported at 2600 K and 82 GPa by Ohta et al. (2015) and up to 100 GPa at 1700 K by Zaghoo et al. (2016), as discussed in Section 6.4.

7.4 Interiors of Uranus and Neptune

The spatial distribution of flux lines of an external magnetic field is affected by the depth below the outer surface of a planet at which its magnetic field is generated, as stated earlier. Historically, the electrical conductivity thought to produce the magnetic fields of U/N was thought to be ionic conductivity of 20/(Ω-cm), measured under adiabatic shock compression for several representative ices up to 75 GPa and 5000 K. At the time of Voyager 2 those magnetic fields of the Ice Giants were thought to be generated near ~0.7 R_U and 0.7 R_N in the ice layers (Nellis et al., 1988a; Podolak et al., 1991). In later years the technique to measure electrical conductivities of fluid H under QI multiple-shock compression was developed (Chapter 6). Subsequent QI dynamic-compression experiments (Weir et al., 1996a; Nellis et al., 1999) imply that the magnetic fields of U/N are probably generated primarily by MFH at radii even closer to their surfaces, perhaps out to ~0.9 $R_{U,N}$ at which pressures in H are ~100 GPa or more and conductivities might be as large as 2000/(Ω-cm).

As shown in Fig. 6.1, electrical conductivity of fluid H at 93 GPa and ~1500 K extends upward more than three orders of magnitude to 2000/(Ω-cm) at 140 GPa, 0.63 g/cm^3 and ~2600 K and remains at ~2000/(Ω-cm) up to 180 GPa (Weir et al., 1996a; Nellis et al., 1999). Those new conductivity measurements of fluid H at 2000/(Ω-cm) supersede the largest conductivities measured previously for ices (Nellis et al., 1988a) by a factor of 10^2. The relevant question to understanding the magnetic fields of U/N then becomes the pressures and likely radii in U/N corresponding to the ~10^2 times larger electrical conductivities of fluid H than thought previously for ices deeper in those planets. Implicit in this question is the assumption that electrical conductivity of fluid H relatively close to the outer surfaces of U/N, at which conductivity is up to 100 times greater than electrical conductivity of 20/(Ω-cm) deeper down in U/N, would generate the dominant contributions to the magnetic fields of U/N.

Gravitational moments measured by Voyager 2 are delocalized constraints, which means they determine the best fit radial-density curve, rather than densities at specific radii. Planetary density is caused by ice, rock, silicates and Fe/Ni, as well as hydrogen. Fig. 7.4 indicates that the crossover to MMC in H occurs over *P/T* intervals of ~50 GPa and ~1000 K, respectively. The core envelop boundaries in U/N are most probably regions of phase crossovers rather than first-order phase transitions. That is, sharp Insulator-metal transitions (IMT) probably do not exist at so-called envelop-core boundaries at finite temperatures in Giant Planets.

A similar estimate of pressures of the crossover is obtained based on measurements at static pressures. At 0.6 g/cm^3 and 300 K, pressure of solid H_2 is ~73 GPa (Loubyere et al., 1996). At ~3000 K solid H_2 is melted and probably dissociated, which means more particles to increase thermal pressure caused by temperatures as high as ~3000 K, which might raise estimated pressure from 73 GPa at 300 K up to ~100 GPa at ~3000 K. Despite the uncertainties in this estimate, the boundaries between envelops and cores in U/N probably occur near ~100 GPa and ~0.6 g/cm^3, the same pressure above which U/N might be composed of pure ice as suggested by Hubbard et al. (1991) based on planetary modeling. This broad region of the crossover spans the bottoms of the envelopes and tops of the cores of U/N. Thus, it is possible that electrical conduction and significant contributions to the magnetic fields of U/N are made at lower radii in envelops and outer radii in the cores of both planets. In contrast to multipolar magnetic fields of U/N probably made near their surfaces at ~0.85 to 0.9 R_U/R_N, the essentially dipolar magnetic field of Earth is made in its convecting fluid-Fe outer core at radii no larger than ~0.5 R_E, where R_E is the radius of Earth.

The fact that the magnetic fields of U/N are probably made close to their outer surfaces offers a qualitative explanation for their complex magnetic-field shapes (Fig. 7.1). A magnetic field can be written as a multi-pole expansion in spherical harmonics. A dipolar magnetic field falls off with distance r from its origin as $1/r^3$. Multipolar fields fall off faster than $1/r^3$ and thus external quadrupolar contributions are more likely to be observed if made relatively close to planetary surfaces. Thus, the observed complex non-dipolar and non-axisymmetric magnetic fields of U/N could be explained simply by the fact that they are made relatively close to their outer surfaces. In this case, no assumption need be made about special interior structures in U/N, as will be discussed.

Magnetic-field morphologies similar to those of U/N have been calculated by Stanley and Bloxham (2004 and 2006). To calculate those magnetic fields, three-dimensional, numerical, MHD dynamo calculations were performed assuming that the source region of the magnetic field is a convecting thin shell surrounding a stably stratified interior. Non-dipolar and non-axisymmetric fields result in part in those MHD calculations from the small length scales imposed by the thin shell. Although such thin-shell dynamos might exist in U/N, there is no known way to determine that thin shell dynamos actually exist there. Their possibility exists but not their certainty.

In addition to deducing a possible explanation for the non-dipolar and non-axisymmetric fields of U/N, it is also important to understand the large tilt angles between the rotational axis and the effective dipolar magnetic axis of U and of

N. Fig. 7.1 suggests that rotational motion of those planets is virtually decoupled from the effective magnetic axis of each planet. In contrast, the Earth's magnetic field is nearly dipolar with its magnetic axis aligned within ~10^0 of its rotational axis. This near alignment of rotational and dipolar axes implies that whole-body rotational motion of Earth is coupled into convective motion of Earth's fluid outer core in which Earth's magnetic field is generated. A clue to finding the cause of the large tilt angles of U/N might lie in the reason why Earth's field is nearly axisymmetric and dipolar.

7.5 Earth's Magnetic Field

Earth's magnetic field is nearly dipolar with its dipolar axis wandering a few degrees over time with respect to its rotational axis. Small deviations of Earth's magnetic field from axial symmetry might be caused by fluctuations in convective flows that generate Earth's magnetic field. Geophysical observations have been made in recent years, which indicate relatively fast (on a geological timescale), intermittent magnetic-field reversals and excursions occur. For example, periodically Earth's magnetic North and South poles are found to reverse in times short compared to times the dipole axis either points essentially north or south (Coe et al., 1995; Sagnotti et al., 2014). In addition, a possible mechanism has been identified for coupling Earth's whole-body rotational motion into convective fluid motion in Earth's fluid outer core (Hide, 1969; Hide et al., 2013) in which Earth's magnetic field is generated. With respect to the latter, if a certain feature of Earth's structure likely produces a given affect on Earth's magnetic field and that feature is missing from U/N, then it is possible that the absent feature enables different magnetic field geometries in U/N relative to Earth.

Because rotational motion (RM) of Earth is apparently coupled into convective dynamo (CD) motions of its fluid-Fe outer core (Fig. 7.1), Earth's RM could stabilize convective motions that generate a dipolar magnetic field. In this case, if a convective fluctuation that tends to destabilize a given dipolar axis occurs, perhaps driven by a giant impact or by some sort of stochastic variation in convective flow patterns, then strong RM-CD coupling either drives convective motions that essentially restore the initial orientation or CD fluctuations that drive the initial magnetic axes out of orientational equilibrium are so strong that RM-CD coupling eventually drives the dipolar axis into an alignment anti-parallel to its initial one.

Understanding these possible scenarios is suggested by the expression for potential energy Φ of the magnetic moment M of a current loop in an effective magnetic field H: $\Phi = -\underline{M} \bullet \underline{H}$, where $\underline{H}$ is essentially the vector sum of local magnetic fields caused by convective current loops in a region. In a magnetic region $\underline{H}$ would be replaced with $\underline{B}$. Suppose further that there is a weak-coupling mechanism between whole-body rotation of Earth and convective motions in the fluid Fe-rich outer core of Earth.

Based on the previously mentioned expression for potential energy Φ, one would expect that over time the fluid outer core would evolve to an ensemble of convective current loops whose number and shapes minimize total magnetic potential energy with $\underline{M}$ and $\underline{H}$ nearly parallel. Such a system is expected to be susceptible to fluctuations in local convective flows, which would cause fluctuations in Earth's nearly dipolar magnetic field in magnitude and orientation, with reversals and significant excursions possible. Dipolar stability of such a dynamo depends on a coupling mechanism that facilitates communication between planetary rotation and convective motions generating the planetary magnetic field.

Interactions between Earth's rotating mantle and fluid outer core have long been considered with respect to their influence on Earth's magnetic field (Hide, 1969; Jacobs, 1975). Recent analysis of deep-Earth seismic signals supports this picture (Hide et al., 2013). Earth has a strong, solid mantle that rotates with nearly constant angular velocity. RM-CD coupling caused tiny variations in the length of a day that were actually observed for 130 years during the era of the British Empire. Possible interaction mechanisms that might cause these variations include surface roughness of ~0.5 km on the inner radius (3500 km) of Earth's strong rock mantle (Hide et al., 2013). This recently discovered possible roughness on the inner surface of Earth's mantle is a possible mechanism coupling planetary rotation and convective conducting Fe flows that make Earth's magnetic field nearly dipolar.

7.6 Magnetic Fields of Uranus and Neptune

In contrast to Earth's strong rock mantle, the envelops of U/N are fluid H_2-He with little strength and large electrical conductivities of fluid H up to as much as ~2000/(Ω-cm). U/N do not have strong solid mantles to couple planetary rotational motion with convective fluid motions in metallic fluid H near outer radii of those planets, at which major contributions to their magnetic fields are probably made.

The cores of U/N are probably heterogeneous complex fluid mixtures at high pressures and temperatures composed of (1) H, O, C, and N formed from decomposed ices and their various reaction products, (2) possible pure metallic fluid H, O, N and their mixtures formed from decomposed ices and remain unreacted and (3) silicates and Fe/Ni that react with decomposed ices and each other. These heterogeneous cores are expected to have numerous nucleation sites for complex convective flows, which suggests numerous active "turbulent" convective cores, which produce relatively small contributions in the cores to the magnetic fields of U/N.

In this picture, magnetic fields are produced primarily by convecting metallic fluid H near the outer radii of U/N, which fields are weakly coupled to magnetic fields produced in the massive heterogeneous cores. In this case, magnetic fields produced primarily by metallic fluid H near the envelop/core interfacial regions might also be expected to be essentially decoupled from rotational motions of the massive cores of U/N, as suggested by large observed tilt angles and magnetic-center offsets in Fig. 7.1.

In this case, the dynamos in U/N that produce the dominant contributions to the external magnetic fields are expected to be relatively free to wander as local convective fluctuations dictate, essentially unconstrained by planetary rotational motion. Thus, tilt angles and center-offsets of the effective dipolar magnetic fields of U/N are expected to vary slowly over the age of U/N. "Polar wander" is probably a better descriptive term for the magnetic fields of U/N than is "polar reversal" for Earth.

7.7 Conclusions

Based on Voyager 2 data and accumulated physical properties of representative planetary fluids measured on Earth at representative *P/T*s in U/N achieved with dynamic compression:

1. Magnetic fields of U/N are probably made primarily by convecting metallic fluid H with measured electrical conductivities as large as 2000/(Ω-cm) (Nellis et al., 1999), 100 times greater than electrical conductivities of 20/(Ω-cm) of likely icy fluids thought previously to make the magnetic fields of U/N by dynamo action (Nellis et al., 1988a). The substantial increase in measured electrical conductivities was made possible by the development of an experimental technique to measure electrical conductivities of fluids under QI compression achieved with multiple shocks, rather than only with single-shock compression, as prior to 1990.
2. Based on measurements of the pressure dependence of conductivities of dense fluid H, metallic fluid H becomes a poor metal at ~0.6 mol H/cm^3, which corresponds to a pressure of ~100 GPa, which is achieved in U/N at radii possibly as large as 0.9 the radii of U/N. The "classical" planetary EOS relating planetary pressure to density places that pressure close to 0 (Fig. 7.3), which is unphysical and incorrect.
3. Because magnetic fields of U/N are made close to their outer surfaces, both quadrupolar and dipolar contributions to the magnetic field are expected, as observed by Voyager 2.
4. Based on comparison of U/N with Earth (Hide et al., 2013), there appears to be no mechanism by which rotational motions of the massive cores of U/N can be

communicated strongly to convective fluid motions that make their magnetic fields by dynamo action primarily in MFH at radii relatively near the surfaces of U/N. In Earth, steady rotation of the strong rock mantle around Earth's rotational axis probably imprints convective motions in the fluid Fe-rich outer core such that potential energy of an ensemble of magnetic moments generated by a corresponding ensemble of convective conducting flows minimizes potential energy Φ of interaction between magnetic moments M and the effective background magnetic field H (or B in a magnetic material) of the fluid Fe outer core.

5. Dynamo convection in the metallic fluid H regions near outer planetary radii is essentially unconstrained and unaffected by planetary rotations of the cores of U/N. The effective dipolar axes of U/N probably wander as local convective fluctuations dictate. Thus, "polar wander" is probably be a better descriptive term for the magnetic fields of U/N than is "polar reversal" for Earth.
6. In this picture, the magnetic fields of U/N would be virtually locked into the metallic H-rich regions near their outer surfaces in which their magnetic fields are primarily made, rather than "frozen" into their cores as has often been thought.

8

Shock-Induced Opacity in Transparent Crystals

To measure temperature from a Planckian spectrum or to collect optical scattering spectra, such as Raman scattering, from a dynamically compressed fluid above ~100 GPa, a transparent electrical insulator with high shock impedance is needed for use as a window/anvil in multiple-shock experiments. As discussed in Chapter 6, for example, MFH has been made at ninefold liquid H_2 density, 140 GPa and calculated 3000 K by reverberating a shock wave in compressible liquid H_2 (ρ_0 = 0.071 g/cm^3) contained between two weakly compressible sapphire crystals (single-crystal Al_2O_3, ρ_0 = 4.0 g/cm^3). It is highly desirable to collect thermal spectra emitted from MFH to measure temperature and to perform Raman scattering experiments to elucidate the nature of MFH.

Unfortunately, optically transparent sapphire crystals at ambient become opaque at shock pressures between 100 GPa and 130 GPa (Urtiew, 1974). Thus, while sapphire is an excellent anvil material mechanically to generate shock reverberation, sapphire is not acceptable as an optical window above $P_H \approx$ ~100 GPa. For this reason a crystal other than sapphire must be used to measure thermal emission and optical scattering spectra, but no shock-compressed high-density transparent crystal has yet been characterized optically for this purpose above 130 GPa.

Shock-induced opacity is generated by crystalline defects induced by fast (supersonic) deformation. Transparent solid insulators are needed at higher shock pressures to open up a newly accessible regime of extreme pressures and temperatures in ultracondensed matter to optical spectrographic measurements. $Gd_3Ga_5O_{12}$ (GGG) with ρ_0 = 7.1 g/cm^3, is a potentially acceptable solution mechanically to this problem. This chapter describes optical experiments on shock-compressed GGG to discover the shock pressure limit for optical temperature measurements and to identify the spectroscopic signature of shock-induced opacity in GGG. Unfortunately, shock-compressed GGG also becomes opaque at 130 GPa, as does sapphire. Nevertheless, those experiments have identified the likely signature of this decades-old problem and imply likely materials that might maintain optical transparency above shock pressures of 130 GPa.

Choice of anvil material affects conditions of *P* and *T* that can be obtained in an ultracompressed fluid. Higher *P* and lower *T* in MFH than achieved to date are achieved with anvil crystals with higher shock impedance $Z = \rho_0 u_s$ than sapphire with $\rho_0 = 4.0$ g/cm^3. GGG was chosen to investigate because it has crystal density 7.1 g/cm^3, which is substantially greater than that of sapphire, and thus has higher shock impedance than sapphire. In addition, an extensive experimental database had already been measured for GGG at shock pressures up to 300 GPa (Mashimo et al., 2006; Zhou et al., 2011; Mao et al., 2011; Zhou et al., 2015; Huang et al., 2015). For completeness, Hugoniot and optical reflectivity data of GGG have been measured in the range 0.3 to 2.7 TPa generated by pulsed high-intensity laser irradiation (Ozaki et al., 2016).

The goal of recent experiments on GGG was to measure thermal emission spectra and radiances to identify quantitatively the shock pressure range in which GGG crosses over from partially transparent to opaque (thermally equilibrated) (Zhou et al., 2015). More generally, the goal was to diagnose the crossover from elastic strong solid at ambient to failure under plastic deformation, to heterogeneous deformation to a weak solid, to a fluid-like solid that equilibrates thermally in the few ns rise time of shock pressure, through melting in bulk on the Hugoniot and, if possible, to a metallic fluid oxide at shock pressures in the melt.

At lower shock pressures in a strong transparent crystal, shock heating is heterogeneous. Effective emission temperatures are a few 1000 K, typical of melting temperatures and a factor of ~10 greater than bulk shock temperatures. Emissivities are low, of order ~0.01. The high temperatures are those of melted grain boundary regions and shear bands, and the low emissivities are caused by a small areal density of emitting hot spots in grain boundaries (Kondo and Ahrens, 1983). At higher shock pressures at which a shocked oxide equilibrates thermally and rapidly in a thin shock front, the emission temperature is expected to be shock temperature of the bulk material. This simple idea was investigated experimentally by measuring shock temperatures of GGG from 40 to 290 GPa.

The range of shock pressures corresponding to those crossovers were diagnosed by fast emission spectroscopy using a 16-channel optical pyrometer with few ns time resolution in the spectral range from 400 to 800 nm. Shock waves were generated with a two-stage light-gas gun at the Institute of Fluid Physics in Mianyang, China (Zhou et al., 2015).

At all shock pressures the respective 16-channel spectra fit gray-body spectra with constant wavelength-independent emissivity ε. There was no noticeable variation in the spectra depending on whether shocked material was deformed heterogeneously or whether the sample was shock-compressed homogeneously in

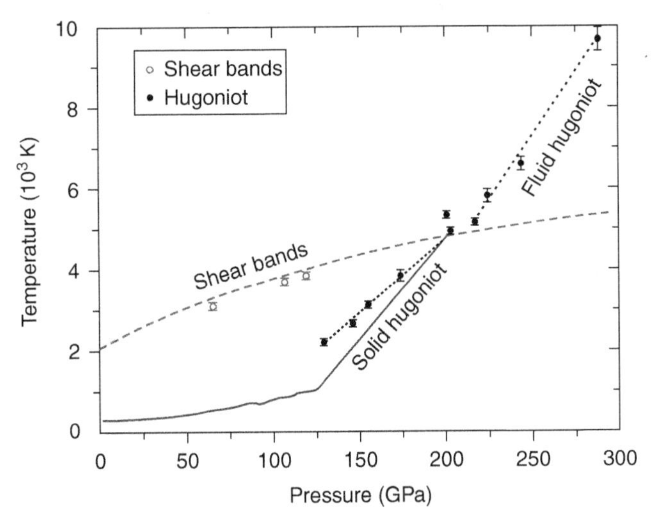

Fig. 8.1. Measured shock temperatures versus pressure for GGG (data points). Experimental lifetimes were few 100 ns. Up to 120 GPa, temperatures are those of shock-induced grain boundaries and shear bands. Between 120 and 130 GPa, temperatures cross over from heterogeneous melting temperatures to thermally equilibrated bulk Hugoniot temperatures. Melting in bulk occurs in two-phase region on Hugoniot from 200 to 220 GPa. Solid curve is calculated Hugoniot of solid phase; dotted curve is calculated Hugoniot of fluid phase. Dashed curve is melting curve calculated with Lindeman model (Zhou et al., 2015). Copyright 2015 by American Institute of Physics.

bulk. The high quality of those 16-channel fits demonstrates that the assumption of wavelength-independent emissivity is an excellent one for thermal radiation from a shock-compressed surface. Measured spectra spanned the shock pressure range from 40 to 290 GPa with corresponding temperatures from 3000 to 8000 K, respectively. Emissivities in those experiments were typically $\varepsilon = 0.5 \pm 0.1$ at shock pressures from 129 to 288 GPa (Zhou et al., 2015).

Fig. 8.1 is a plot of measured temperatures versus shock pressures determined by shock impedance matching the impacts. Below 130 GPa, thermal emission is heterogeneous and measured emission temperatures are dominated by melting at pressures in slip planes and grain-boundary regions. At 130 GPa and 2200 K, GGG equilibrates thermally in bulk on the Hugoniot, as indicated by the shock pressure at which measured temperature achieves near equality with shock temperature calculated with a Gruneisen model, which assumes thermal equilibrium.

Spectral radiances measured from 65 to 129 GPa are plotted versus time in Fig. 8.2. Opacity is caused by optical scattering and absorption of light at defects induced by fast compression. GGG melts at ~5000 K in a two-phase region at shock pressures in the range 200 to 220 GPa (Zhou et al., 2015). Extrapolation of

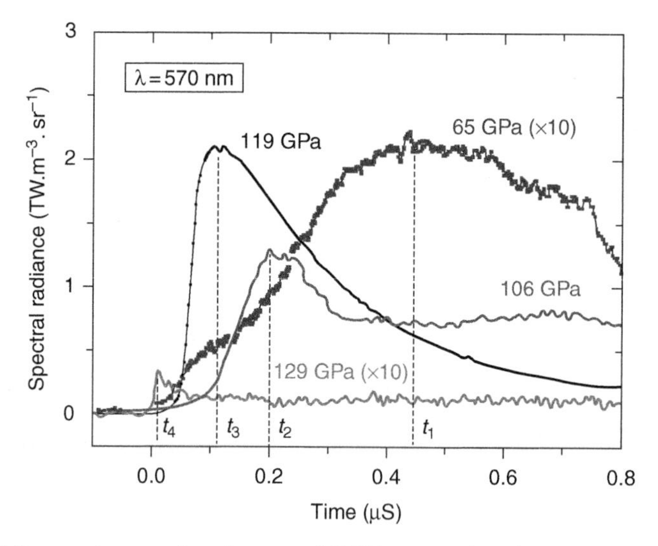

Fig. 8.2. Measured spectral radiances of GGG versus time in 570 nm channels at pressures between 65 and 129 GPa. Vertical dashed lines are at times of peak radiances of respective curves. Curves for 65 and 129 GPa are magnified by factor of 10 for clarity. Shock-compressed GGG is heated heterogeneously and partially transparent below 120 GPa and opaque and thermally equilibrated above 129 GPa (Zhou et al., 2015). Copyright 2015 by American Institute of Physics.

electrical conductivity data measured below ~260 GPa (Mashimo et al., 2006) suggests GGG becomes a poor metal with MMC at shock pressures above 400 GPa, as is observed in laser-driven shock experiments and associated theoretical calculations (Ozaki et al., 2016).

Because GGG goes opaque at 130 GPa, as does sapphire, it is probably not possible to measure temperatures of MFH made with GGG anvils. However, because GGG has higher shock impedance than sapphire, it might be possible to measure lower temperatures at higher pressures than in the semiconducting fluid phase of hydrogen made with sapphire. Such measurements with GGG might provide an experimental test of the EOS model used to calculate densities and temperatures in dense fluid H.

The result that strong GGG and sapphire both go opaque by shock-induced defects at 130 GPa implies the possibility that an anvil material with lower density and higher compressibility might be used to measure temperature without going opaque until shock pressures are substantially higher than 130 GPa. One such possibility is LiF with initial density 2.7 g/cm^3, which is reported to be transparent up to a shock pressure of 200 GPa (Furnish et al., 1999).

9

Metastable Solid Metallic Hydrogen (MSMH)

MFH has been made under dynamic compression essentially at predicted metallization density of solid H, 0.62 mol H/cm^3 (Wigner and Huntington, 1935). If MFH could be quenched to metastable solid metallic H (MSMH) at ambient, then several conceivable technological and scientific applications with substantial potential benefits to science and society might become available. The classical theoretical prediction in this regard is that MSMH at density 0.62 mol/cm^3 and ninefold density of liquid H_2 at ambient pressure is a room-temperature superconductor (Ashcroft, 1968), which implies substantial energy efficiencies might be achieved in long-distance electrical power transmission, for example. However, because equilibrium density of condensed H_2 near ambient pressure is so much smaller than that of MFH, quenching MFH to MSMH without significant decrease in density is a grand challenge for science and technology.

Other potential technological applications of MSMH are based on its high atomic density, rather than electrical conductivity. These include fuels and propellants, energy-storage media and lightweight strong and ductile structural materials (Nellis, 1999). These potential uses of MSMH require appropriate mechanical properties of dense hydrogen and hydrogenous materials, but without the need to have H density as high as needed to make a metal, which is expected to facilitate making metastable nonmetallic hydrogenous materials.

Scientific applications include materials science of making unusual materials at extreme conditions with the aid of theoretical computations to guide/interpret small-scale experiments. Addressing key scientific materials issues in small-scale dynamic compression experiments has the advantage that small-scale dynamic experiments using two stage guns and lasers are efficient for developing techniques that can be scaled up in size using pressures generated by large high explosive-driven dynamic systems, for example, as has been done at E. I. DuPont de Nemours & Company with ceramic powders (Bergmann and Barrington, 1966).

Given the enormous potential benefits to society, the attempt to make MSMH is certainly worth a substantial effort (Nellis, 1999, 2006).

The quest for metallic hydrogen has been going on for well over 100 years. This quest began in the 1890s with searches for cryogenic liquids with low-temperature boiling and freezing points. H_2 was first liquefied and then solidified in 1898 and 1899, respectively, by James Dewar. Prior to those accomplishments it was expected that condensed phases of hydrogen near atmospheric pressure would be monatomic alkali metals, as elements below H in the first column of the Periodic Table (Mendelssohn, 1966). Instead, condensed phases of hydrogen turned out to be transparent diatomic insulators, as halogens in the seventh column of the Periodic Table. Something more than intuition was needed to figure out why liquid and solid hydrogen are transparent insulators rather than opaque metals. That something turned out to be quantum mechanics. And so, the long quest for metallic solid hydrogen began at the end of the nineteenth century and has continued into the twenty-first.

After the development of quantum mechanics in the 1920s WH predicted (1935) that body-centered cubic (*bcc*) H_2 would undergo a first-order transition to *bcc* metallic H at density 0.62 mol H/cm^3, very low temperatures T and some unimaginably large pressure $P > 25$ GPa, recently reported to be 495 GPa (Dias and Silvera, 2017). Mott (1936) pointed out that a very low temperature of an electron system is one that is degenerate, that is, $T/T_F << 1$ where T is temperature of the electron system and T_F its Fermi temperature, which depends on density at 0 K. This requirement is likely achieved with dynamic compression because both density and temperature increase under dynamic compression and so T_F increases as T increases. That is, ultracondensed degenerate matter is readily achieved with dynamic compression in many cases.

MSMH has been predicted to be a high-temperature superconductor in both monatomic H (Ashcroft, 1968) and diatomic H_2 (Richardson and Ashcroft, 1997) crystalline phases. Jaffe and Ashcroft (1981) predicted that liquid metallic H might be a superconductor at essentially the same high pressure and density as predicted for solid metallic H ($1.6 > r_s > 1.3$) with $T_c \approx 100$ K. In 1996, MFH was discovered under dynamic compression at 140 GPa, 0.63 mol H/cm^3 ($r_s \approx 1.6$) and $T \approx 2600$ K. Subsequently, the melting temperature T_m of solid hydrogen at density $r_s \approx 1.6$ and 140 GPa was calculated and observed to be $T_m \approx 800$ K (Bonev et al., 2004; Deemyad and Silvera, 2008).

Because T_m of solid H_2 is much greater than the predicted T_c of liquid metallic H (~100 K), that T_c prediction for the liquid is mute up to quite high pressures. To check if MFH is a superconductor at T_c ~ 100 K, it is necessary to quench MFH to solid amorphous H for cryogenic testing. However, it is reasonable to expect the possibility that, if liquid metallic H is predicted to be superconductor, then amorphous (quenched-liquid) H might also be a superconductor. Whether liquid or amorphous metal, the key point is that metallic bonding of monatomic H with itinerant electrons is essential for a metallic superconductor, rather than bonding with electrons localized in H_2 diatomic bonds. Thus, there is a possibility that the predicted metallic nature and superconductivity of liquid metallic H might be observed in metastable amorphous metallic H (MAMH), if it could be made.

Many amorphous systems are known to superconduct. Amorphous Bi and Ga become superconducting at $T_c \sim 6.0$ K and ~8.5 K, respectively (Chen et al., 1969). Bulk crystalline Bi has not been observed to superconduct and crystalline Ga superconducts at $T_c = 1.1$ K, much less than that of the amorphous Ga. The crystalline intermetallic compound $LuRh_4B_4$ superconducts at Tc ~ 7 K. Disorder induced by radiation damage by alpha particles decreases its normal-state electrical conductivity to minimum metallic conductivity of 2500/(ohm-cm) and superconductivity persists in the disordered solid phase at $T_c = 1$ K (Dynes et al., 1981).

To increase the probability of making MFH and quenching it to metastable amorphous metallic H (MAMH), it would probably be best if higher densities and lower temperatures were achieved in MFH prior to initiation of the quench. Higher densities (lower r_s) might increase the range of stability of the superconducting state, and lower temperatures in MFH would facilitate thermal quenching. To accomplish higher ρ and lower T, anvils used to make MFH (Fig. 6.1) should have higher density than sapphire (Al_2O_3), which would produce a greater shock-impedance mismatch with liquid H_2. Thus, the probability of success of quenching a superconducting state might be increased by increasing the density of MFH from the expected maximum allowed value of $r_s \approx 1.6$ down to $r_s \approx 1.5$ to 1.4. Higher densities in MFH might be achieved if Al_2O_3 anvils with density 4.00 g/cm^3 were replaced with an oxide such as GGG with density 7.10 g/cm^3 (Zhou et al., 2015), if possible.

Because of the high diatomic bond energy of the H_2 molecule 4.5 eV, when dynamic pressure on MFH releases in a quench, charge neutrality will almost certainly drive recombination of H atoms back to electrically insulating H_2 molecules and associated ambient density of liquid H_2. This recombination is the main impediment to making MAH. Finding a technique to retain high densities of atomic H on release of dynamic high pressures is the key problem. Physical chemistry techniques at high shock pressures (Dlott, 2011) might be very useful in the quest to identify a useful technique to retain MAH after release of MFH from high pressures.

One approach might be to mix additive(s) with a hydrogenous sample compressed dynamically such that the additives would react chemically with hydrogen at high dynamic P/T to form bonds sufficiently strong to maintain metallic densities achieved at high pressures on release of pressure to ambient. However, some amount of "free" (conduction) electrons would need to be retained on release to achieve a poor metal that could superconduct. An unusual material might be made if such a process could be successfully devised. It is important that mixtures be uniform initially on an atomic scale for rapid reaction. Dynamic experimental lifetimes are probably too brief to depend on atomic diffusion to form an appropriate mixture.

The release process might also be tuned to occur in stages to avoid a single large release to ambient. Because pressures might extend up to the ~100 GPa range, computational design of materials and experimental testing would be required. Achievement of metastable amorphous solid H (MASH) is a formidable goal that requires substantial multidisciplinary expertise and efforts in physics, chemistry, materials science and hydrodynamics, but if successful, such a project has the potential to change life as we know it.

MFH has a density of ~0.6 g/cm^3 at high *P/T*. With additives to make it quenchable to MSH with appropriate mechanical properties of strength, ductility, and so forth for various applications, density of materials recovered from high pressures might be, say, ~1–2 g/cm^3, about the density of water or somewhat greater. Such materials might be useful for (1) a source of H_2 for H fuel cells; (2) lightweight structural materials for autos and so forth; (3) fuel and propellant for autos and space travel, depending on rate of energy release; (4) a room-temperature superconductor for energy-efficient electrical power transmission; and (5) dense D-T fuel pellets for higher energy yields in ICF (Nuckolls et al., 1972; Motz, 1979).

A limiting phenomenon of ICF is growth of turbulent mixing between D-T fuel pellet and its encapsulation material. If encapsulation material is mixed with D-T fuel, thermal energy (heating) deposited in D-T under substantial dynamic compression would partially be absorbed in ionizing electrons in the fuel capsule rather than heating the D-T. Such ionization would lower the temperature T of the D-T, and the fusion burn rate R of D-T is proportional to ~T^4. As a result, R would decrease substantially relative to what it would be in the absence of this loss. Such mixing by interfacial instability is minimized by matching mass densities of D-T fuel and encapsulation material (Section 2.2.2). Dense MSM(D-T) fuel, in contrast to gaseous or liquid D-T, would facilitate matching those densities, thus facilitating substantial ICF fusion-energy yields.

In a manner analogous to the way fossil fuels drove the worldwide industrial revolution of the twentieth century, ICF is positioned to perform an analogous role in the Twenty-First century. Of course, development of ICF on a global commercial scale is contingent on the requisite scientific and technological research and development.

In the 1950s, Enrico Fermi said about nuclear physics: "History of science and technology has consistently taught us that scientific advances in basic understanding have sooner or later led to technical and industrial applications that have revolutionized our way of life." Such a program on MMSH would keep ICF's scientific and engineering research vital, as well as potentially create massive employment in new energy and other industries previously referred to. ICF might provide the commercial energy required to drive those technological applications. ICF is an international issue and requires international commitments analogous to those at the Large Hadron Collider at the European Center for Nuclear Research (CERN) in Switzerland.

10

Warm Dense Matter at Shock Pressures up to 20 TPa (200 Mbar)

Warm dense matter (WDM) is atomic matter for which potential energy of electron-ion interactions is comparable to kinetic energy of electrons. WDM has densities typically in the range between 0.01 and 100 g/cm^3 and temperatures that range up a few hundred thousands of Kelvins and higher, generally in the range between 1 and 100 eV (1 eV = 1.16×10^4 K) (Landau and Zeldovich, 1943; Norman et al., 2013; Saitov, 2016) as illustrated in Fig. 10.1. In this regime WDM encompasses ionized fluids at the confluence of condensed matter physics, plasma physics and dense fluids, which exist in ICF and in cores of extrasolar giant planets. A comprehensive review of matter at extreme conditions has been published (Fortov, 2016). In this chapter systematics of a universal crossover from ultracondensed fluids, including melted solids, to WDM along the R-H curve are discussed.

Ultracondensed matter at finite T is degenerate "cold" matter at ultrahigh densities. Plasma matter is non-degenerate "hot" matter. To characterize finite temperatures T at a given density ρ, T increases from cold to warm to hot depending on degeneracy factor T/T_F, where T_F is Fermi temperature of the Fermi-Dirac electron distribution. T_F is density at 0 K expressed as the characteristic temperature of the Fermi-Diarac distribution. In this context, cold, warm and hot are defined broadly in terms of $T/T_F << 1$, $T/T_F \approx 1$ and $T/T_F >> 1$, respectively.

Condensed matter has been investigated extensively under shock and multiple-shock compression up to a few 100 GPa (Marsh, 1980; Trunin, 1998 and 2001; Nellis, 2006a). Those experimental data have been used to generate relatively simple EOSs of solids and liquids that depend on (P, V, E). The most common such EOS model is the Gruneisen model, which is a reasonable EOS approximation provided the Grueisen parameter $\gamma = V(\partial P/\partial E)_V$depends only on density, independent of temperature (McQueen et al., 1970). Thermal pressure P_T is pressure generated by T at a given V. In the Gruneisen model $P = P_{ref}$ (V,T) + γE,

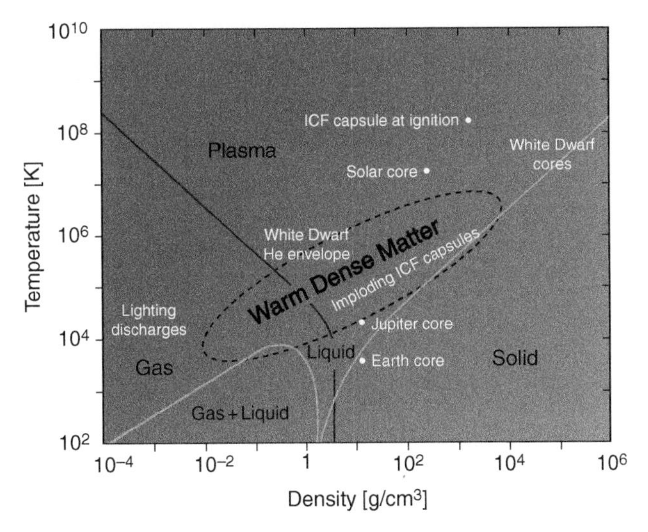

Fig. 10.1. Densities and temperatures of warm dense matter (WDM) and plasma. WDM is ionized fluids at the confluence of condensed matter physics, plasma physics and dense liquids. Copyright 2011 by Los Alamos National Security, LLC 2016

where P is total pressure, $P_{ref}(V,T)$ is a reference pressure and γE is thermal pressure generated by internal energy deposited by compression.

The Gruneisen model is commonly used up to 100–200 GPa to describe material under shock compression. γ is generally calculated theoretically or derived from measured double-shock EOS data that have substantial uncertainties. In this case the R-H curve is P_{ref} (V,T). Measured or theoretically calculated 300-K isotherms are also used as P_{ref} (V,T). The Gruneisen EOS is commonly used in hydrodynamic simulations as described in Section 2.1.16 (Nellis, 2006a, pp. 1498–1500).

Use of the Grueisen EOS to simulate off-Hugoniot EOS states is limited to maximum compressions of ~threefold ρ_0 because maximum shock compression is ~fourfold ρ_0. Up to ~100–200 GPa, thermal pressure is generally less than a few 10% of Hugoniot pressure at a given V. In genera,l Gruneisen EOSs differ with materials at these pressures (Chijioke et al., 2005a). In addition dynamic strength, phase transitions (Duvall and Graham, 1977) and phase crossovers are observed for many materials at these pressures. Those phenomena do not have a common effect on experimental EOS data. Thus, although the Gruneisen EOS is a common form up to ~200 GPa, a common universal EOS has not been observed for condensed matter up to a few 100 GPa.

Dynamic compression experiments today are performed with giant pulsed lasers and hypervelocity impacts generated by magnetically driven metal plates. These experiments are crossing over to higher pressures from condensed matter to the WDM regime, which is more complicated than condensed matter physics probed previously. However, the complexity and frequency of techniques now being used are such that substantially fewer experiments are performed than previously at lower dynamic pressures.

Experimental data for WDM are needed for comparison with evolving theories of matter in this regime. Strong dynamic compression achieves high pressures and temperatures well above melting in the fluid phase, which causes materials to enter the WDM regime. First-order phase transitions are smeared out in WDM because of shock-induced temperatures and disorder. Crossovers exist over relatively wide ranges of P, V, T with no obvious sharp features to constrain theory.

However, an extensive Hugoniot database of a wide range of WDM materials was measured between 1970 and 1990 in a wide range of shock pressures from 0.3 TPa (300 GPa) up to 20 TPa (200 Mbar) and higher (1) located near underground nuclear explosions (Ragan, 1984; Mitchell et al., 1991; Trunin, 2001), (2) generated by hypervelocity impacts at the Z Accelerator (Knudson and Desjarlais, 2009) and (3) generated with a high-intensity pulsed laser (Bradley et al., 2004; Ozaki et al., 2016). An early search for a signature of systematic behavior in WDM made by strong shock compression found a possible indication that not only are there no sharp features characteristic of WDM in Hugoniot data of individual materials but also, remarkably, measured shock-compression data in shock-velocity/particle-velocity space of several WDM materials might lie on essentially the same featureless straight line – essentially independent of material, shock pressure and shock temperature. This shock-velocity/particle-velocity curve is essentially a universal Hugoniot for a large number of materials. That possibility (Nellis, 2006b) motivated a recent study (Ozaki et al., 2016). In this case the experimental feature to test theoretical developments is answering the question as to why does WDM have a universal Hugoniot weakly dependent on material, pressure, and temperature. This observation was totally unexpected.

10.1 Analysis of Published Hugoniot Data from 0.3 to 20 TPa

As mentioned, an extensive database of measured Hugoniot EOS data for WDM has been accumulated over the past fifty years. Those experiments produced (P_H, V_H, E_H) data on Hugoniots of materials that are interesting both in themselves and for generating systematic results as a possible guide for understanding WDM. Remarkably, those experimental data demonstrate unexpected systematic behavior, which begs the question as to why dynamic compression produces systematically similar results for such a wide variety of materials and thermodynamic conditions.

In the search for systematic behavior, the first thing to try is simply plotting measured data in terms of variables that might indicate some sort of universal behavior. An optimal place to search for common "universal" behavior is to consider "scaled data", data that are independent of a common parameter that differs substantially from material to material. For example, Hugoniots of diatomic fluids H_2, N_2, O_2 and CO are nearly identical in $u_s(u_p)$ space (Fig. 2.14), even though those Hugoniots differ dramatically in $P(V)$ space (Nellis, 2002b).

A preliminary test of this idea was made previously for fluid metals Al, Cu, Fe and Mo above ~0.5 TPa. Published Hugoniot points in $u_s(u_p)$ space of these metals were least-squares fit to $u_s = C + Su_p$, with $C = 5.90$ km/s and $S = 1.22$ in the range $4 < u_p < 42$ km/s (Nellis, 2006b). Shock pressure corresponding to the point at 42 km/s is 20 TPa (200 Mbar). Because this $u_s(u_p)$ fit was determined by least-square fitting measured data for several fluid metals over an enormous range of shock pressures, we call this fit the Universal Hugoniot of fluid metals (UHFM).

The previous data analysis was undertaken to determine if curvature expected in Hugoniot data induced by TPa shock pressures and associated high temperatures would be observed as thermally induced electronic excitation, as observed, for example, in fluid Ar above a shock pressure of 50 GPa (Chapter 5). Those Ar excitations are caused by absorption of shock-induced thermal energy, which excites valence electrons across a mobility gap of ~10 eV into the conduction band of dense fluid Ar. A similar analysis of published "underground data" failed to detect effects of thermal activation in available $u_s(u_p)$ data measured up to 20 TPa (Nellis, 2006b).

More recently, analysis of previously measured Hugoniot EOS data of metals at extreme conditions was expanded to forty published points of fluid Al, Mo, Fe and Cu at shock pressures in the range from 0.3 to 20 TPa, with estimated corresponding shock temperatures ranging up to 10^6 K. Thirty-three points for the four metals were measured with the shock-impedance match method using several assumed shock-pressure standards (Trunin, 2001). Seven Al points were measured in impact experiments in which impact velocity and shock velocity in the Al samples were measured (Knudson et al., 2003b).

Fig. 10.2 is a $u_s(u_p)$ plot of those forty measured Hugoniot points with the previously determined UHFM (Nellis, 2006b) superimposed. The linear UHFM fit to the new larger data set is $u_s = C + Su_p$, with $C = 5.8$ km/s and $S = 1.2$ in the range $5 < u_p < 45$ km/s (Ozaki, 2015), which is essentially identical to the fit to the previous smaller data set.

For a free-electron gas, $S = 1.33$ (Nellis, 2003), which implies that $S = 1.2$ (< 1.33) might be caused by electron correlations in WDM. There is no

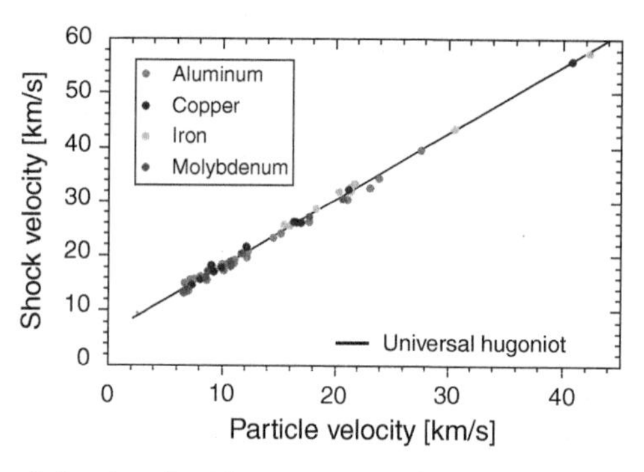

Fig. 10.2. Measured shock velocities versus particle velocities for Al, Mo, Fe and Cu at corresponding shock pressures from 0.3 to 20 TPa (200 Mbar). Measured Hugoniot points were generated with underground nuclear explosions, chemical explosions and Z Accelerator. Hugoniot data measured for GGG with pulsed laser and Z Accelerator also lie on solid line known as Universal Hugoniot of fluid metals (UHFM) (Ozaki et al., 2016). Copyright 2016 Nature-Scientific Reports.

indication of shell-structure effects nor of a crossover from WDM to plasma on the common Hugoniot below 20 TPa. Shell-structure effects are most likely to occur in those metals at shock pressures above 20 TPa and temperatures above ~10^6 K (Rozsnyai et al., 2001), which is consistent with these UHFM data fits in the sense that shell-structure effects are not observed in $u_s(u_p)$ space below 20 TPa.

Shell-structure effects in Al are predicted theoretically at shock pressures from 40 to 400 TPa (Rozsnyai et al., 2001). Unfortunately, error bars on measured data at those ultrahigh shock pressures are too large to verify the existence of those effects. However, those measurements at 100 TPa shock pressures are substantial scientific accomplishments, given the difficulty of the environment in which they were made.

The term UHFM probably includes all materials, whether a metal at ambient (Mo, Fe, Cu, etc.) or an insulator at ambient ($Gd_3Ga_5O_{12}$, SiO_2 quartz, rare gas liquids, etc.) that become metallic at sufficiently high shock pressures and temperatures. To test this idea, Hugoniot data of GGG were measured at pressures in the range 0.4 to 2.63 TPa generated with a pulsed laser at Osaka University and by impact at the Z Accelerator. Results of both GGG experiments are in good agreement with each other and with the UHFM (Ozaki et al., 2016). The $u_s(u_p)$ fit of SiO_2 (quartz) was measured previously up to 1.6 TPa (Knudson and Desjarlais, 2009) and is essentially identical to the UHFM.

Measured Hugoniots at shock pressures in the range 350 to 860 GPa of liquids Xe and Kr with initial densities 2.97 g/cm^3 and 2.43 g/cm^3, respectively, have linear $u_s(u_p)$ relations with $C = 1.624$ km/s and $S = 1.163$ for Xe (Root et al., 2010) and $C = 1.313$ km/s and $S = 1.231$ for Kr (Mattesson et al., 2014). These S values of rare-gas fluids are similar to those of the UHFM. These values of C are lower than that of the UHFM with a higher average initial density than rare gas liquids, which is not unexpected.

Ahuja and colleagues have calculated that disordered GGG becomes a poor metal with optical reflectivity of ~0.15 at ~1.2 TPa on its Hugoniot, in agreement with measured optical reflectivities (Ozaki et al., 2016). The reflectivity calculations were performed by computationally quenching a GGG crystal from 8000 K from various initial densities to an amorphous state. The shock pressure corresponding to that of the quenched sample was calculated using the density assumed in a given calculation and the $u_s(u_p)$ fit of the UHFM (Fig. 10.2), which is in good agreement with measured GGG Hugoniot data (Ozaki et al., 2016).

10.2 Measured and Calculated Optical Reflectivities of GGG above 0.4 TPa

Fig. 10.3 is a plot of measured optical reflectivities at wavelength 0.532 nm of a GGG shock front versus shock pressures determined from measured shock-velocity history of the front of a decaying shock wave (Fig. 2.11). For comparison, also shown are reflectivities calculated at corresponding shock pressures calculated as previously described. Agreement between measured and calculated reflectivities

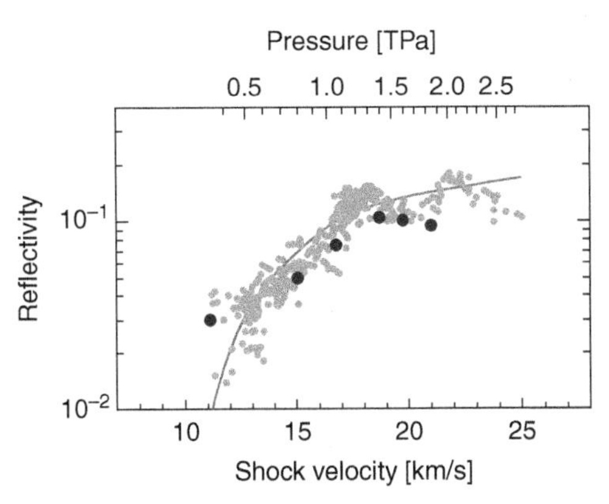

Fig. 10.3. Measured optical reflectivities at wavelength 532 nm (gray dots) from front of decaying shock wave in GGG as functions of measured shock velocity and corresponding calculated shock pressures. Shown for comparison are theoretically calculated optical reflectivities (dark solid circles) versus corresponding shock pressures of shock-melted wave front as described in text (Ozaki et al., 2016). Copyright 2016 Nature-Scientific Reports.

is excellent. An optical reflectivity of 0.15 is typical of a disordered poor metal and corresponds to MMC of a few $10^3/\Omega$-cm.

10.3 Universal State of Ultracondensed Matter and WDM: Atomic Fluids with MMC

Fluid metallic degenerate H (Nellis et al., 1999), O (Bastea et al., 2001) and N (Chau et al., 2003a) have MMC, as do fluid Rb and Cs (Hensel and Edwards, 1996; Hensel et al., 1998) and Li (Bastea and Bastea, 2002). MMC is commonly observed for liquid metals under QI multiple-shock compression and heated under static compression. MMC is caused by strong electron scattering with mean-free scattering length comparable to inter-atomic distances in disordered materials (Mott, 1972).

Above a shock pressure of ~TPa, the Hugoniot of fluid GGG is coincident with UHFM and has an optical reflectivity of ~0.15, which is a typical reflectivity of a material with MMC. The measured Hugoniot of fluid SiO_2 quartz (Knudson and Desjarlais, 2009) lies on the UHFM from 0.7 to 1.6 TPa. Calculated low-frequency electrical conductivities of fluid SiO_2 range from 1000 to 6000/(Ω-cm) as shock pressure increases from 0.1 to 1 TPa with corresponding temperatures ranging from 8000 to 76,000 K. Theoretical structural calculations indicate fluid SiO_2 dissociates virtually completely in this same range of shock pressures (Laudernet et al., 2004). These results indicate fluid SiO_2 on its Hugoniot has MMC in that range of shock pressure with electron mean-free path comparable to inter-atomic distances. These results for GGG and quartz imply that materials with UHFM have MMC. The calculated conductivity of shock-melted Al_2O_3 on its Hugoniot at 0.9 TPa is also MMC (Liu et al., 2015a). All of these results indicate electrical insulators, such as strong oxides, become poor metals at sufficiently high shock pressures and temperatures.

Fig. 10.2 suggests that in the range from ~0.3 to 20 TPa the Hugoniots of most fluids of materials with initial crystal densities between 2.5 and 11.5 g/cm^3 probably lie on the UHFM. In the experiments in Fig. 10.2, densities of those fluid metals determined by Trunin and his colleagues (1998) are 5.0 to 9.0 g/cm^3, 22 to 27 g/cm^3, 15 to 30 g/cm^3 and 18 to 34 g/cm^3 for Al, Mo, Fe and Cu, respectively. Corresponding temperatures range up to an estimated several 10^5 K. Since those elements are metals at ambient, they are expected to remain metallic and have MMC at the extreme pressures and temperatures on the UHFM.

Evidence is mounting that it is quite likely that all fluids reach MMC at sufficiently extreme pressures and temperatures in a "universal" state of dense "atomic" matter at finite temperatures, a state typified by strong disorder scattering that produces electrical conductivities of MMC. This state is observed under QI multiple-shock compression, under adiabatic single-shock compression and under heated static compression. It is likely, based on compression experiments to date, that this state is independent of the way it is produced, provided pressure and temperature are sufficiently large.

Fig. 10.2 raises some fundamental questions about the nature of thermodynamics and pair distribution functions achieved in dense fluids by supersonic hydrodynamics:

1. Why is it that Hugoniots of materials above ~0.5 TPa have linear $u_s(u_p)$ relations?
2. (a) Why is it that those linear $u_s(u_p)$ relations are all quite likely coincident, at least for materials with initial densities primarily between ~2.5 g/cm^3 and ~10. 5 g/cm^3? That is, what is it about shock compression that apparently causes an effectively common, universal, linear $u_s(u_p)$ relation over such a wide range?
   (b) Why in particular are the C and S coefficients of the UHFM, given by C = 5.8 km/s and S = 1.2? That is, why is it that apparently all atoms over an enormous range of pressures and temperatures apparently know to fall on essentially the same $u_s(u_p)$ line?

Electron correlations might be part of the answer to these questions.

10.4 Warm Dense Matter Analogue of Asymptotic Freedom of High Energy Physics

Asymptotic freedom is a property of quantum chromodynamics (QCD), a theory for nuclear interactions between quarks and gluons, the fundamental building blocks of nuclear matter. QCD allows bonds between particles to become asymptotically weaker as energy increases with decreasing interparticle distance (Gross and Wilczek, 1973; Politzer, 1973). In this section we propose that the crossover from strong electrically insulating metal-oxygen chemical bonds to a compressible metallic fluid with hybridized electron energy bands is an analogue of asymptotic freedom. This crossover on the GGG Hugoniot from strong condensed matter to compressible WDM occurs between 0.3 and 1 TPa via a bonds-to-bands crossover.

Shock compression data for GGG up to 0.25 TPa (Mashimo et al., 2006) and from 1.0 to 2.6 TPa (Ozaki et al., 2016) are plotted in Fig. 10.4. In the lower-pressure range, GGG is a strong oxide that is less compressible than diamond from 0.17 to 0.26 TPa on the Hugoniot. However, in the range from 1.0 to 2.6 TPa, shock compression of GGG is ~30% greater than expected from extrapolation of the lower-pressure data up to shock pressures of ~1 TPa. Low compressibility of GGG below 0.25 TPa is probably caused by the existence of strong metal-oxygen

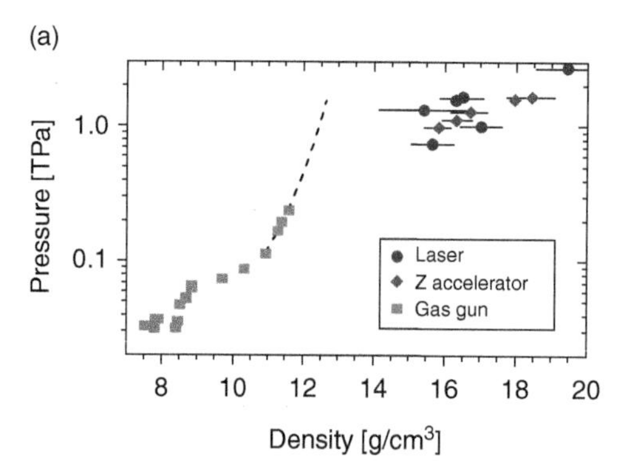

Fig. 10.4. Measured Hugoniot pressures versus densities of GGG. Low compressibility of data at shock pressures near ~0.3 TPa compared to data with high compressibility at shock pressures near ~1 TPa implies GGG under shock compression is a condensed matter analogue of asymptotic freedom in quark matter in high-energy particle physics (Ozaki et al., 2016). Copyright 2016 Nature-Scientific Reports.

bonds. At 0.07 and 0.25 TPa on the Hugoniot, GGG is a semiconductor with electrical conductivities 10^{-2} and $10^{1}/(\Omega-\text{cm})$, respectively (Mashimo et al., 2006). In contrast, above ~1.0 TPa the Hugoniot of fluid GGG is in good agreement with the UHFM and has an optical reflectivity of ~0.15 (Fig 10.3), which is typical of a poor metal with MMC of a few $10^{3}/(\Omega\text{-cm})$. The crossover of GGG from condensed matter to WDM occurs in the same range of shock pressures in which optical reflectivity of the GGG shock front increases from negligible to ~0.15. WDM on the UHFM is expected to be a metal with MMC. The coincidence of the crossover from condensed matter to UHFM with the crossover from semiconductor to poor metal confirms that these crossovers are a signature of the crossover from condensed matter to WDM.

References

Ahrens, T. J. (2005). Shock wave experiments. In *Encyclopedia of Geomagnetism and Paleomagnetism*. ed. D. Gubbins. Dordrecht: Kluwer, pp. 912–920.

Akahama, Y. and Kawamura, H. (2007). Diamond anvil Raman gauge in multimegabar pressure range. *High Pressure Research*, **27**, 473–482.

Alexander, C. S., Asay, J. R. and Haill, T. A. (2010). Magnetically applied pressure-shear: A new method for direct measurement of strength at high pressure. *Journal of Applied Physics*, **108**, 126101-1–126101-3.

Altshuler, L. V. (1965). Use of shock waves in high-pressure physics. *Soviet Physics-Uspekhi*, **8**, 52–91.

Altshuler, L. V. (2001). *Lost World of Khariton*. Sarov: RCNC-VNIIEF.

Altshuler, L. V., Kormer, S. B., Brazhnik, M. I., Vladimirov, L. A., Speranskaya, M. P. and Funtikov, A. I. (1960). The isentropic compressibility of aluminum, copper, lead and iron at high pressures. *Soviet Physics – JETP*, **11**, 766–776.

Altshuler, L. V., Podurets, M. A., Simakov, G. V. and Trunin, R. F. (1973). High-density forms of fluorite and rutile. *Soviet Physics Solid State*, **15**, 969–971.

Altshuler, L. V., Trunin, R. F., Krupnikov, K. K. and Panov. N. V. (1996). Explosive laboratory devices for shock wave compression studies. *Physics-Uspekhi*, **39**, 539–544.

Altshuler, L. V. et al. (1968). *Soviet Journal of Experimental and Theoretical Physics*, **54**, 785–789.

Anderson, M. S. and Swenson, C. A. (1974). Experimental compressions for normal hydrogen and normal deuterium to 25 kbar at 4.2 K. *Physical Review B*, **10**, 5184–5191.

Ao, T., Asay, J. R., Chantrenne, S., Baer, M. R. and Hall, C. A. (2008). A compact strip-line pulsed power generator for isentropic compression experiments. *Review of Scientific Instruments*, **79**, 013903-1–013903-16.

Arnett, D. (1996). *Supernovae and Nucleosynthesis* (First edition). Princeton: Princeton University Press.

Asay, J. R., Chhabildas, L. C., Lawrence, R. J. and Sweeney, M. A. (2017). *Impactful Times: Memories of 60 Years of Shock Wave Research at Sandia National Laboratories*. Cham, Switzerland: Springer International Publishing AG.

Asay, J. R. and Shahinpoor, M. (Eds.). (1993). *High Pressure Shock Compression of Solids*. New York: Springer.

Ashcroft, N. W. (1968). Metallic hydrogen: A high-temperature superconductor? *Physical Review Letters*, **21**, 1748–1749.

Ashcroft, N. W. (1991). Optical response near a band overlap: application to dense hydrogen. In *Molecular Systems under High Pressure*, eds. R. Pucci and G. Piccitto. New York: Elsevier, pp. 201–222.

Babaev, E., Sudbo, A. and Ashcroft, N. W. (2004). A superconductor to superfluid phase transition in liquid metallic hydrogen. *Nature*, **431**, 666–668.

Bancroft, D., Peterson, E. L. and Minshall, S. (1956). Polymorphism of iron at high pressure. *Journal of Applied Physics*, **27**, 291–298.

Banishev, A. A., Shaw, W. L., Bassett, W. P. and Dlott, D. D. (2016). High-speed laser-launched flyer plate impacts studied with ultrafast strobe photography and photon Doppler velocimetry. *International Journal of Impact Engineering*. DOI 10.1007/s40870-016-0058-2.

Barker, L. M. (1984). High pressure quasi-isentropic experiments. In *Shock Waves in Condensed Matter-1983*, eds. J. R. Asay, R. A. Graham and G. K. Straub. New York: North-Holland, pp. 217–224.

Barker, L. M. and Hollenbach, R. E. (1974). Shock wave study of the $\alpha \Leftrightarrow \varepsilon$ phase transition in iron. *Journal of Applied Physics*, **45**, 4872–4887.

Barnes, J. F., Blewett, P. J., McQueen, R. G. and Meyer, K. A. (1974). Taylor instability in solids. *Journal Applied Physics*, **45**, 727–732.

Bassett, W. A. (2009). Diamond anvil cell, 50th birthday. *High Pressure Research*, **29**, 163–186.

Bastea, M. and Bastea, S. (2002). Electrical conductivities of lithium at megabar pressures. *Physical Review B*, **65**, 193104-1–193104-4.

Bastea, M., Mitchell, A. C. and Nellis, W. J. (2001). High-pressure insulator-metal transition in molecular fluid oxygen. *Physical Review Letters*, **86**, 3108–3111.

Bean, V. E., Akimoto, S., Bell, P. M., Block, S., Holzapfel, W. B., et al. (1982). Toward an international practical pressure scale: an AIRAPT task group report. In *High Pressure in Research and Industry*, vol. 1, eds. C. M. Backman, T. Johannisson and L. Tegner. Uppsala: Arkitektkopia, pp. 144–151.

Bean, V. E., Akimoto, S., Bell, P. M., Block, S., Holzapfel, W. B., et al. (1986). Another step toward an international practical pressure scale: 2nd AIRAPT IPPS Task Group Report. *Physica*, **139/140B**, 52–54.

Belov, S. I., Boriskov, G. V., Bykov, A. I., Il'kaev, R. I., Luk'yanov, N. B., et al. (2002). Shock compression of solid deuterium. *JETP Letters*, **76**, 433–435.

Benson, D. J. and Nellis, W. J. (1994). Dynamic compaction of copper powder: computation and experiment. *Applied Physics Letters*, **65**, 418–420.

Bergmann, O. R. and Barrington, J. (1966). Effect of explosive shock waves on ceramic powders. *Journal of the American Ceramic Society*, **49**, 502–507.

Bethe, H. A. and Teller, E. 1940. *Deviations from Thermal Equilibrium in Shock Waves*. Aberdeen Ballistics Research Laboratory, unpublished report. (Harvard University Library QC718.5.W3B56).

Bethe, H. A. 1942. *On the Theory of Shock Waves for an Arbitrary Equation of State*. U. S. Department of Commerce, Washington, DC, Technical Report PB-32189.

Boehler, R., Ross, M., Soderlind, P. and Boercker, D. (2001). High-pressure melting curves of argon, krypton, and xenon: deviation from corresponding states theory. *Physical Review Letters*, **86**, 5731–5734.

Bonev, S. A., Schwegler, E., Ogitsu, T. and Galli, G. (2004). A quantum fluid of metallic hydrogen suggested by first-principles calculations. *Nature* (London), **431**, 669–672.

Boriskov, G. V., Bykov, A. I., Il'kaev, R. I., Selemir, V. D., Simakov, G. V., et al. (2005). Shock compression of liquid deuterium up to 109 GPa. *Physical Review B*, **71**, 092104-1–092104-4.

Bowden, S. A., Parnell, J. and Burchell, M. J. (2009). Survival of organic compounds in ejecta from hypervelocity impacts on ice. *International Journal of Astrobiology*, **8**, 19–25.

Bradley, D. K., Eggert, J. H., Hicks, D. G., Celliers, P. M., Moon, S. J., et al. (2004). Shock compressing diamond to a conducting fluid. *Physical Review Letters*, **93**, 195506-1–195506-4.

Bridgman, P. W. (1909). The measurement of high hydrostatic pressure, I. A simple pressure gauge. *Proceedings of the American Academy of Arts and Sciences*, **44**, 201–217.

Bridgman, P. W. (1935). Theoretically interesting aspects of high pressure phenomena. *Reviews of Modern Physics*, **7**, 1–33.

Bridgman, P. W. (1956). High pressure polymorphism of iron. *Journal of Applied Physics*, **27**, 659.

Bridgman, P. W. (1959). Compression and the $\alpha-\beta$ phases transition of plutonium. *Journal of Applied Physics*, **30**, 214–217.

Bridgman P. W. (1963). In *Solids Under Pressure*, ed. W. Paul and D. M. Warschauer, NY: McGraw-Hill, pp. 1–13.

Bridgman, P. W. (1964). *Collected Experimental Papers* (seven volumes). Cambridge: Harvard University Press.

Bushman A. V., Kanel G. I., Fortov V. E. and Ni A. L. (1992). *Intense Dynamic Loading of Condensed Matter*. Bristol: Taylor and Francis.

Calder, A. C., Fryxell, B., Plewa, T., Rosner, R. et al. (2002). On validating an astrophysical simulation code. *Astrophysical Journal Supplement Series*, **143**, 201–229.

Cavazonni, C., Chiarotti, G. L., Scandolo, S., Tosatti, E., et al. (1999). Superionic and metallic states of water and ammonia at giant planet conditions. *Science*, **283**, 44–46.

Celliers, P. M., Collins, G. W., Da Silva, L. B., Gold, D. M., Cauble, R., et al. (2000). Shock-induced transformation of liquid deuterium into a metallic fluid. *Physical Review Letters*, **84**, 5564–5567.

Challis, J. (1848). On the velocity of sound. *Philosophical Magazine*, **32**, 494–499.

Charters, A. C., Denardo, B. P. and Rossow, V. J. (1957). Development of a piston-compressor type light-gas gun for the launching of free-flight models at high velocity. National Advisory Committee for Aeronautics Technical Note 4143.

Chau, R., Mitchell, A. C., Minich, R. and Nellis, W. J. (2001). Electrical conductivity of water compressed dynamically to pressures of 70–180 GPa (0.7–1.8 Mbar). *Journal of Chemical Physics*, **114**, 1361–1365.

Chau, R., Mitchell, A. C., Minich, R. W. and Nellis, W. J. (2003a). Metallization of fluid nitrogen and the Mott transition in highly compressed low-Z fluids. *Physical Review Letters*, **90**, 245501-1–245501-4.

Chau, R., Bastea, M., Mitchell, A. C., Minich, R. W. and Nellis, W. J. (2003b). Planetary interior in the laboratory. *Lawrence Livermore National Laboratory Report*. UCRL-ID-151657.

Chau, R., Hamel, S. and Nellis, W. J. (2011). Chemical processes in the deep interior of Uranus. *Nature Communications*, **2**, 203. DOI: 10.1038/ncomms1198.

Chen, Q. F., Zheng, J., Gu, Y. J., Chen, Y. L., Cai, L. C. and Shen Z. J. (2014). Thermophysical properties of multi-shock compressed dense argon. *Journal of Chemical Physics*, 140, 074202-1–074202-6.

Chen, Q. F., Zheng, J., Gu, Y. J. and Li, Z. G. (2015). Equation of state of dense neon and krypton plasmas in the partial ionization regime. *Physics of Plasmas*, 22, 122706-1–122706-8.

Chen, T. T., Chen, J. T., Leslie, J. D. and Smith, H. J. T. (1969). Phonon spectrum of superconducting amorphous bismuth and gallium by electron tunneling. *Physical Review Letters*, **22**, 526–530.

Cheret, R. (1992). The life and work of Pierre-Henri Hugoniot. *Shock Waves*, **2**, 1–4.

Cheret, R. (1993). *Detonation of Condensed Explosives*. New York: Springer.

Chijioke, A. D., Nellis, W. J., and Silvera, I. F. (2005a). High-pressure equations of state of Al, Cu, Ta, and W. *Journal of Applied Physics*, **98**, 073526-1–073526-8.

Chijioke, A. D., Nellis, W. J., Soldatov, A. and Silvera, I. F. (2005b). The ruby pressure standard to 150 GPa. *Journal of Applied Physics*, **98**, 114905-1-114905-9.

Coe, R. S., Prevot, M. and Camps, P. (1995). New evidence for extraordinarily rapid change of the geomagnetic field reversal. *Nature*, 374, 687–692.

Collins, L., Kress, J., Kwon, I., Windl, W., Lenosky, T., et al. (1998). Quantum molecular dynamics simulations of dense matter. *Journal of Computer-Aided Materials Design*, **5**, 173–191.

Courant, R. and Friedrichs, K. O. (1948). *Supersonic Flow and Shock Waves*. New York: Springer-Verlag, pp. 1–3.

Cronin, J. W. (2004). Fermi, E., The future of nuclear physics. In *Fermi Remembered*. Chicago: University of Chicago Press.

Dalladay-Simpson, P., Howie, R. T. and Gregoryanz, E. (2016). Evidence for a new phase of dense hydrogen above 325 gigapascals. *Nature*, **529**, 63–67.

Davis, J.-P., Brown, J. L., Knudson, M. D. and Lemke, R. W. (2014). Analysis of shockless dynamic compression data on solids to multi-megabar pressures: Application to tantalum. *Journal of Applied Physics*, **116**, 204903-1–204903-17.

Davison, L. and Graham, R. A. (1979). Shock compression of solids. *Physics Reports*, **55**, 255–379.

Decker, D. L., Bassett, W. A., Merrill, L., Hall, H. T. and Barnett, J. D. (1972). High-pressure calibration: A critical review. *Journal of Physical and Chemical Reference Data*, **1**, 1–79.

Deemyad, S. and Silvera, I. F. (2008). Melting line of hydrogen at high pressures. *Physical Review Letters*, **100**, 155701-1–155701-4.

Dias, R. and Silvera, I. F. (2017). Observation of the Wigner-Huntington transition to solid metallic hydrogen. *Science*, *355*, 715–718.

Dick, R. D. and Kerley, G. I. (1980). Shock compression data for liquids. II. Condensed hydrogen and deuterium. *Journal of Chemical Physics*, **73**, 5264–5271.

Dlott, D. D. (2011). New developments in the physical chemistry of shock compression. *Annual Review of Physical Chemistry*, **62**, 575–597.

Dorfman, S. M., Jiang, F., Mao, Z., Kubo, A., Meng, Y., et al. (2010). Phase transitions and equations of state of alkaline earth fluorides CaF_2, SrF_2, and BaF_2 to Mbar pressures. *Physical Review B*, **81**, 174121-1–174121-13.

Dorogokupets, P. I., Sokolova, T. S., Daniliov, B. S., Litasov, K. D. (2012). Near-absolute equations of state of diamond, Ag, Al, Au, Cu, Mo, Nb, Pt, Ta, and W for quasi-hydrostatic conditions. *Geodynamics & Tectonophysics*, **3**, 129–166. DOI:10.5800/GT-2012-3-2-0067.

Drickhamer, H. G. and Balchan, A. S. (1961). High pressure electrical resistance cell, and calibration points above 100 kilobars. *Review of Scientific Instruments*, **32**, 308–313.

Dubrovinskaia, N., Dubrovinsky, L., Kantor, I., Crichton, W. A., Dmitriev, V., et al. (2005). Beating the miscibility barrier between iron group elements and magnesium by high-pressure alloying. *Physical Review Letters*, **95**, 245502-1–245502-4.

Dubrovinsky, L., Dubrovinskaia, N., Prakapenka, V. B. and Abakumov, A. M. (2012). Implementation of micro-ball nanodiamond anvils for high-pressure studies above 6 Mbar. *Nature Communications*, **3**, 1163-1–1163-7.

Duvall, G. E. and Graham, R. A. (1977). Phase transitions under shock-wave loading. *Reviews of Modern Physics*, **49**, 523–579.

Dynes, R. C., Rowell, J. M. and Schmidt, P. H. (1981). In *Ternary Superconductors* ed. G. K. Shenoy, B. D. Dunlap and F. Y. Fradin. Amsterdam: North Holland, pp. 169–173.

Dzyabura, V., Zaghoo, M. and Silvera, I. F. (2013). Evidence of a liquid-liquid phase transition in hot dense hydrogen. *Proceedings of National Academy of Sciences (U.S.A)*, **110**, 8040–8044.

Eakins, D. E. and Thadhani, N. N. (2009). Shock compression of reactive powder mixtures. *International Materials Review*, **54**, 181–213.

Earnshaw, S. (1860). On the mathematical theory of sound. *Philosophical Transactions of the Royal Society of London*, **150**, 133–148.

Edwards, P. P., Johnston, R. L., Rao, C. N. R., Tunstall, D. P. and Hensel, F. (1998). The metal-insulator transition: a perspective. *Philosophical Transactions of the Royal Society of London A*, **356**, 5–22.

Eggert, J., Brygoo, S., Loubeyre, P., McWilliams, R. S., Celliers, P. M., et al. (2008). Hugoniot data for helium in the ionization regime, *Physical Review Letters*, **100**, 124503-1–124503-4.

Eggert, J. H., Hicks, D. G., Celliers, P. M., Bradley, D. K., McWilliams, R. S., et al. (2009). Melting temperature of diamond at ultrahigh pressure. *Nature Physics*, **6**, 40–43.

Eremets, M. I. (1996). *High Pressure Experimental Methods*. Oxford: Oxford University Press.

Erskine, D. J. and Nellis, W. J. (1991). Shock-induced martensitic phase transformation of oriented graphite to diamond. *Nature*, **349**, 317–319.

Erskine, D. J. and Nellis, W. J. (1992). Shock-induced martensitic transformation of highly oriented graphite to diamond. *Journal of Applied Physics*, **71**, 4882–4886.

Erskine, D. (1994). High pressure Hugoniot of sapphire. In *High Pressure Science and Technology – 1993*, eds. S. C. Schmidt, J. W. Shaner, G. A. Samara and M. Ross. New York: American Institute of Physics Conference Proceedings 309, pp. 141–143.

Feynman, R. P. (1963). *Six Easy Pieces*. New York: Basic Books,.

Forbes, J. W. (2012). *Shock Wave Compression of Condensed Matter: A Primer (Shock Wave and High Pressure Phenomena)*. Heidelberg: Springer-Verlag.

Fortov, V. E., Ternovoi, V. A., Zhernokletov, M. V., Mochalov, M. A., Mikhailov, A. L., et al. (2003). Pressure-produced ionization of nonideal plasma in a megabar range of dynamic pressures. *Journal of Experimental and Theoretical Physics*, **97**, 259–278.

Fortov, V. E., Altshuler, L. V., Trunin, R. F. and Funtikov, A. I., eds. (2004). *High-Pressure Shock Compression of Solids VII, Shock Waves and Extreme States of Matter*. New York: Springer-Verlag.

Fortov, V. E. (2016). *Extreme States of matter: High energy density physics* (Second edition). New York: Springer.

Furakawa, Y., Sekine, T., Oba, M., Kakegawa, T. and Nakazawa, H. (2009). Biomolecule formation by oceanic impacts on early Earth. *Nature Geoscience*, **2**, 62–66.

Furnish, M. D., Chhabildas, J. C. and Reinhart, W. G. (1999). Time-resolved particle velocity measurements at impact velocities of 10 km/s. *International Journal of Impact Engineering*, **23**, 261–270.

Gatilov, L. A., Glukhodedov, V. D., Grigorev, F. V., Kormer, S.B., Kuleshova, L. V. and Mochalov, M. A. (1985). Electrical conductivity of shock compressed condensed argon at pressures from 20 to 70 GPa. *Journal of Applied Mechanics and Technical Physics*, **26**, 88–91.

Goncharov, A. F., Goldman, N., Fried, L. E., Crowhurst, J. C., Kuo, I. W., Mundy, C. J. and J. M. Zaug. (2005). Dynamic ionization of water under extreme conditions. *Physical Review Letters*, **94**, 125508-1–125508-4.

Gorman, M. G., Briggs, R., McBride, E. E., Higginbotham, A., Arnold, B., et al. (2015). Direct observation of melting in shock-compressed bismuth with femtosecond X-ray diffraction. *Physical Review Letters*, **115**, 095701-1–095701-5.

Grady, D. E. (1998). Shock-wave compression of brittle solids. *Mechanics of Materials*, **29**, 181–203.

Graham, R. A. (1993). *Solids Under High Pressure Shock Compression*. New York: Springer.

Graham, R. A. (1994). Bridgman's concern. In *High Pressure Science and Technology – 1993*, eds. S. C. Schmidt, J. W. Shaner, G. A. Samara and M. Ross. New York: American Institute of Physics Conference Proceedings 309, pp. 3–12.

Granzow, K. D. (1983). Spherical harmonic representation of the magnetic field in the presence of a current density. *Geophysical Journal of the Royal Astronomical Society*, **74**, 489–505.

Gratz, A. J., Nellis, W. J. and Hinsey, N. A. (1993a). Observations of high-velocity, weakly shocked ejecta from experimental impacts. *Nature*, **363**, 522–524.

Gratz, A. J., DeLoach, L. D., Clough, T. M. and Nellis, W. J. (1993b). Shock amorphization of cristobalite. *Science*, **259**, 663–666.

Grigoryev, F. B., Kormer, S. B., Mikhailova, O. L., Mochalov, M. A. and Urlin, V. D. (1985). Temperatures of shock compressed liquid nitrogen and argon. *Journal of Experimental and Theoretical Physics*, **88**, 1271–1280.

Gross, D. J. and Wilczek, F. (1973). Ultraviolet behavior of non-abelian gauge theories. *Physical Review Letters*, **30**, 1343–1346.

Hall, C. A., Asay, J. R., Knudson, M. D., Stygar, W. A., Spielman, R. B. and Pointon, T. D. (2001). Experimental configuration for isentropic compression of solids using pulsed magnetic loading. *Review of Scientific Instruments*, **72**, 3587–3595.

Hamann, S. D. and Linton, M. (1966). Electrical conductivity of water in shock compression. *Transactions of the Faraday Society*, **62**, 2234–2241.

Hamilton, D. C., Nellis, W. J., Mitchell, A. C., Ree, F. H. and van Thiel, M. (1988a). Electrical conductivity and equation of state of shock compressed liquid oxygen. *Journal of Chemical Physics*, **88**, 5042–5050.

Hamilton, D. C., Mitchell, A. C., Ree, F. H. and Nellis, W. J. (1988b). Equation of state of 1-butene shocked to 54 GPa (540 kbars). *Journal of Chemical Physics*, **88**, 7706–7708.

Hare, D. E., Holmes, N. C. and Webb, D. J. (2002). Shock-wave-induced optical emission from sapphire in the stress range 12 to 45 GPa: Images and spectra. *Physical Review B*, **66**, 014108-1–014108-11.

Hawke, R. S., Burgess, T. J., Duerre, D. E., Huebel, J. G., Keeler, R. N., et al. (1978). Observation of electrical conductivity of isentropically compressed hydrogen at megabar pressures. *Physical Review Letters*, **41**, 994–997.

Helled, R., Anderson, J. D., Podolak, M. and Schubert, G. (2011). Interior models of Uranus and Neptune. *Astrophysical Journal*, 726, 15-1–15-7.

Hemley, R. J., Chiarotti, G. L., Bernasconi, M. and Ulivi, L. (2002). *High Pressure Phenomena: Proceedings of the International School of Physics "Enrico Fermi."* Amsterdam: IOS Press.

Hensel, F. and Edwards, P. P. (1996). The changing phase of expanded metals. *Physics World*, **4**, 43–46.

Hensel, F., Marceca, E. and Pilgrim, W. C. (1998). The metal-non-metal transition in compressed metal vapours. *Journal of Physics: Condensed Matter*, 10, 11395–11404.

Herman, F. and Skillman, S. (1963). *Atomic Structure Calculations*. Englewood Cliffs, NJ: Prentice-Hall.

Hicks, D. G., Celliers, P. M., Collins, G. W., Eggert, J. H. and Moon, S. J. (2003). Shock-induced transformation of Al_2O_3 and LiF into semiconducting liquids. *Physical Review Letters*, **91**, 035502-1–035502-4.

Hicks, D. G., Boehly, T. R., Celliers, P. M., Eggert, J. H., Moon, S. J., et al. (2009). Laser-driven single shock compression of fluid deuterium from 45 to 220 GPa. *Physical Review B*, **79**, 014112-1–014112-18.

Hide, R. (1969). Interaction between the Earth's liquid core and solid mantle. *Nature*, **222**, 1055–1056.

Hide, R., Clayton, R. W., Hager, B. H., Spieth, M. A. and Voorhdes, C. V. (2013). Topographic core-mantle coupling and fluctuations in the Earth's rotation. In *Relating Geophysical Structures and Processes: The Jeffreys Volume, Geophysical Monograph 76*, Vol. 16, eds. K. Aki and R. Dmowska, IUGG, pp. 107–120.

Holme, R. (1997). Three dimensional kinematic dynamos with equatorial symmetry: Application to the magnetic fields of Uranus and Neptune. *Physics of the Earth and Planetary Interiors*, **102**, 105–122.

Holmes, N. C., Nellis, W. J., Graham, W. B. and Walrafen, G. E. (1985). Spontaneous Raman scattering from shocked water. *Physical Review Letters*, **55**, 2433–2436.

Holmes, N. C., Moriarty, J. A., Gathers G. R. and Nellis, W. J. (1989). The equation of state of platinum to 660 GPa (6.6 Mbar). *Journal of Applied Physics*, **66**, 2962–2967.

Holmes, N. C., Ross, M. and Nellis, W. J. (1995). Temperature measurements and dissociation of shock-compressed liquid deuterium and hydrogen. *Physical Review B*, **52**, 15835–15845.

Holmes, N. C., Nellis, W. J. and Ross, M. (1998). Sound velocities in shocked liquid deuterium. In *Shock Compression of Condensed Matter – 97*, eds. S. C. Schmidt, D. P. Dandekar, and J. W. Forbes. Woodbury, NY: American Institute of Physics Conference Proceedings 429, pp. 61–64.

Holst, B., French, M. and Redmer, R. (2011). Electronic transport coefficients from *ab initio* simulations and application to dense liquid hydrogen. *Physical Review B*, **83**, 235120-1–235120-8.

Holzapfel, W. B. (2005). Progress in the realization of a practical pressure scale for the range 1–300 GPa. *High Pressures Research*, **25**, 87–96.

Holzapfel, W. B. (2010). Equations of state for Cu, Ag, and Au and problems with shock wave reduced isotherms. *High Pressures Research*, **30**, 372–394.

Hoover, W. G. (1979). Structure of a shock-wave front in a liquid. *Physical Review Letters*, **42**, 1531–1534.

Horie, Y. and Sawaoka, A. B. (1993). *Shock Compression Chemistry of Materials*. Tokyo: KTK Scientific Publishers.

Hu, S. X., Goncharov, V. N., Boehly, T. R., McCrory, R. L., Skupsky, S., et al. (2015). Impact of first-principles properties of deuterium-tritium on inertial confinement fusion target designs. *Physics of Plasmas*, **22**, 056304-1–056304-14.

Huang, J. W., Liu, Q. C., Zeng, X. L., Zhou, X. M. and Luo, S. N. (2015). Refractive indices of $Gd_3Ga_5O_{12}$ single crystals under shock compression to 100–290 GPa. *Journal of Applied Physics*, **118**, 205902-1–205902-4.

Hubbard, W. B. (1981). Interiors of the Giant Planets. *Science*, **214**, 145–149.

Hubbard, W. B. (1984). *Planetary Interiors*. New York: Van Nostrand-Rheinhold, New York.

Hubbard, W. B., Nellis, W. J., Mitchell, A. C., Holmes, N. C., et al. (1991). Interior structure of Neptune: Comparison with Uranus. *Science*, **253**, 648–651.

Hubbard, W. B., Podolak, M. and Stevenson, D. J. (1995). The interior of Neptune. In *Neptune and Triton*, eds. D. P. Cruikshank. Tuscon: University of Arizona Press, pp. 109–138.

Hugoniot, H. (1887). Sur la propagation du mouvement dans les corps et spécialement dans les gaz parfaits-I. *Journal of Ecole Polytechnique*, **157**, 3–98.

Hugoniot, H. (1889). Sur la propagation du mouvement dans les corps et spécialement dans les gaz parfaits-II. *Journal Ecole Polytechnique*, **158**, 1–126.

Hurricane, O. A., Callahan, D. A., Casey, D. T., Celliers, P. M., et al. (2014). Fuel gain exceeding unity in an inertially confined fusion implosion. *Nature*, **506**, 343–348.

Inoue, K., Kanzaki, H. and Suga, S. (1979). Fundamental absorption spectra of solid hydrogen. *Solid State Communications*, **30**, 627–629.

Ioffe, A. F. and Regal, A. R. (1960). Non-crystalline, amorphous, and liquid electronic semiconductors. *Progress in Semiconductors*, **4**, 237–291.

Irwin, P. G. J. (2003). *Giant Planets of Our Solar System: Atmospheres, Composition, and Structure*. New York: Springer-Verlag.

Isbell, W. M. (2005). *Shock Waves: Measuring Dynamic Response of Materials*. London: Imperial College Press.

Jacobs, J. A. (1975). *The Earth's Core*. London: Academic Press.

Jaffe, J. E. and Ashcroft, N. W. (1981). Superconductivity in liquid metallic hydrogen. *Physical Review B*, **23**, 6176–6179.

Jamieson, J. C. and Lawson, A. W. (1962). X-ray diffraction studies in the 100 kilobar pressure range. *Journal of Applied Physics*, **33**, 776–780.

Jayaraman, A. (1983). Diamond anvil cell and high-pressure physical investigations. *Reviews of Modern Physics*, **55**, 65–108.

Jayaraman, A. (1984). The diamond-anvil high-pressure cell. *Scientific American*, April, 54–63.

Jones, A. H., Isbell, W. M. and Maiden, C. J. (1966). Measurement of the very-high-pressure properties of materials using a light-gas gun. *Journal of Applied Physics*, **37**, 3493–3499.

Kalantar, D. H., Belak, J. F., Collins, G. W., Colvin, J. D., Davies, M. H., et al. (2005). Direct observation of the $\alpha-\varepsilon$ transition in shock-compressed iron via nanosecond X-ray diffraction. *Physical Review Letters*, **95**, 075502-1–075502-4.

Kanel, G. I., Razorenov, S. V. and Fortov, V. E. (2004). *Shock-Wave Phenomena and the Properties of Condensed Matter*. New York: Springer.

Kanel, G. I., Nellis, W. J., Savinykh, A. S., Razorenov, S. V. and Rajendran, A. M. (2009). Response of seven crystallographic orientations of sapphire crystals to shock stresses of 16–86 GPa. *Journal of Applied Physics*, **106**, 043524-1–043524-10.

Kaxiras, E., Broughton, J. and Hemley, R. J. (1991). Onset of metallization and related transitions in solid hydrogen. *Physical Review Letters*, **67**, 1138–1141.

Kerley, G. I. (1983). A model for the calculation of thermodynamic properties of a fluid. In *Molecular-Based Study of Fluids*, eds. J. M. Haile and G. A. Mansoori. Washington DC: American Chemical Society, pp. 107–138.

Klimenko, V. Y. and Dremin, A. N. (1979). In *Detonatsiya, Chernogolovka*, eds. O. N. Breusov et al. Moscow: Soviet Academy of Sciences, p. 79.

Knudson, M. D., Hanson, D. L., Bailey, J. E., Hall, C. A., Asay, J. R. and Anderson, W. W. (2001). Equation of state measurements in liquid deuterium to 70 GPa. *Physical Review Letters*, **87**, 255501-1–255501-4.

Knudson, M. D., Hanson, D. L., Bailey, J. E., Hall, C. A. and Asay, J. R. (2003a). Use of a wave reverberation technique to infer the density compression of shocked liquid deuterium to 75 GPa. *Physical Review Letters*, **90**, 035505-1–035505-4.

Knudson, M. D., Lemke, R. W., Hayes, D. B., Hall, C. A., Deeney, C. and Asay, J. R. (2003b). Near-absolute Hugoniot measurements in aluminum to 500 GPa using a magnetically accelerated flyer plate technique. *Journal of Applied Physics*, **94**, 4420–4431.

Knudson, M. D., Hanson, D. L., Bailey, J. E., Hall, C. A., Asay, J. R. and Deeney, C. (2004). Principal Hugoniot, reverberating wave, and mechanical reshock measurements of liquid deuterium to 400 GPa using plate impact techniques. *Physical Review B*, **69**, 144209-1–144209-20.

Knudson, M. D., Asay, J. R. and Deeney, C. (2005). Adiabatic release measurements in aluminum from 240- to 500-GPa states on the principal Hugoniot. *Journal of Applied Physics*, **97**, 073514-1–073514-14.

Knudson, M. D. and Desjarlais, M. P. (2009). Shock compression of Quartz to 1.6 TPa: Redefining a pressure standard. *Physical Review Letters*, **103**, 255501-1–255501-4.

Knudson, M. D. and Desjarlais, M. P. (2013). Adiabatic release measurements in α-quartz between 300 and 1200 GPa: Characterization of α-quartz as a standard in the multi-megabar regime. *Physical Review B*, **88**, 184107-1–184107-18.

Knudson, M. D., Desjarlais, M. P., Becker, A., Lemke, R. W., Cochrane, K. R., et al. (2015). Direct observation of an abrupt insulator-to-metal transition in dense liquid deuterium. *Science*, **348**, 1455–1459.

Kondo, K. and Ahrens, T. J. (1983) Heterogeneous shock-induced thermal radiation in minerals. *Physics and Chemistry of Minerals*, **9**, 173–181.

Kormer, S. B. (1968). Optical study of the characteristics of shock-compressed condensed dielectrics. *Soviet Physics-Uspekhi*, **11**, 229–254.

Kraus, R. G., Senft, L. E. and Stewart, S. T. (2011). Impact onto H2O ice: Scaling laws for melting, vaporization and final crater size. *Icarus*, 214, 724–738.

Kraus, R. G., Stewart, S. T., Swift, D. C., Bolme, C. A., Smith, R. F., et al. (2012). Shock vaporization of silica and the thermodynamics of planetary impact events. *Journal of Geophysical Research*, **117**, E09009-1–E09009-22.

Kraus, R. G., Root, S., Lemke, R. W., Stewart, S. T., Jacobsen, S. B. and Mattsson, T. R. (2015). Impact vaporization of planetesimal cores in the late stages of planet formation. *Nature Geoscience*, 8, 269–272.

Kraus, D., Ravasio, A., Gauthier, M., Gericke, D. O., et al. (2016). Nanosecond formation of diamond and lonsdaleite by shock compression of graphite. *Nature Communications*, **7**, DOI: 10.1038/ncomms10970.

Krehl, P. O. K. (2015). The classical Rankine-Hugoniot jump conditions, an important cornerstone of modern shock wave physics: Ideal assumptions vs. reality. *European Physical Journal H*, **40**, 159–204.

Kuhlbrodt, S., Redmer, R., Reinholz, H., Ropke, G., Holst, B., et al. (2005). Electrical conductivity of Noble gases at high pressures. *Contributions to Plasma Physics*, **45**, 61–69.

Kunc, K., Loa, I. and Syassen, K. (2003). Equation of state and phonon frequency calculations of diamond at high pressures. *Physical Review B*, **68**, 094107-1–094107-9.

Landau, L. D. and Zeldovich, Y. B. (1943). On the relation between the liquid and the gaseous states of metals. *Acta Physico-Chimica USSR*, **18**, 174.

Laudernet, Y., Clerouin, J. and Mazevet, S. (2004). *Ab initio* simulations of the electrical and optical properties of shock-compressed SiO_2. *Physical Review B*, **70**, 165108-1–165108-5.

Lee, K. K. M., Benedetti, L. R., Jeanloz, R., Celliers, P., Eggert, J. H., et al. (2006) Laser-driven shock experiments on precompressed water: Implications for "icy" giant planets. *Journal of Chemical Physics*, **125**, 014701-1–014701-7.

Lemke, R. W., Knudson, M. D. and Davis, J-P. (2011). Magnetically driven hyper-velocity launch capability at the Sandia Z accelerator. *International Journal of Impact Engineering*, **38**, 480–485.

Lin, F., Morales, M. A., Delaney, K. T., Peirleoni, C., Martin, R. M. and Ceperley, D. M. (2009). Electrical conductivity of high-pressure liquid hydrogen by Quantum Monte Carlo methods. *Physical Review Letters*, **103**, 256401-1–256401-4.

Lindl, J., Landen, O., Edwards, J., Moses, E. and NIC Team. (2014). Review of the National Ignition Campaign 2009–2012. *Physics of Plasmas*, **21**, 020501-1–020501-72.

Liu, T.-P. (1986). Shock waves for compressible Navier-Stokes equations are stable. *Communications on Pure and Applied Mathematics*, **39**, 565–594.

Liu, H., Tse, J. S. and Nellis, W. J. (2015). The electrical conductivity of Al2O3 under shock compression. *Scientific Reports*, 5, 12823-1–12823-9.

Liu, Q., Zhou, X., Zeng, X. and Luo, S. N. (2015). Sound velocity, equation of state, temperature and melting of LiF single crystals under shock compression. *Journal of Applied Physics*, 117, 045901-1–045901-6.

Lorenzen, W., Holst, B. and Redmer, R. (2010). First-order liquid-liquid phase transition in dense hydrogen. *Physical Review B*, **82**, 195107-1–195107-6.

Loubeyre, P., LeToullec, R., Hausermann, D., Hanfland, M., Hemley, R. J., et al. (1996). X-ray diffraction and equation of state of hydrogen at megabar pressures. *Nature (London)*, **383**, 702–704.

Lucas, M., Winey, J. M. and Gupta, Y. M. (2015). Sound velocities in highly oriented pyrolytic graphite shocked to 18 GPa: Orientational order dependence and elastic instability. *Journal of Applied Physics*, **118**, 245903-1–245903-9.

Luo, S.-N., Akins, J. A. and Ahrens, T. J. (2004). Shock-compressed MgSiO3 glass, enstatite, olivine, and quartz: Optical emission, temperatures, and melting. *Journal of Geophysical Research*, 109, B05205-1–B05205-14.

Lyzenga, G. A. and Ahrens, T. J. (1982). One-dimensional isentropic compression. In *Shock Waves in Condensed Matter – 1981*, eds. W. J. Nellis, L. Seaman and R. A. Graham. New York: American Institute of Physics Conference Proceedings 78, pp. 231–235.

Lyzenga, G. A., Ahrens, T. J., Nellis, W. J. and Mitchell, A. C. (1982). The temperature of shock-compressed water. *Journal of Chemical Physics*, **76**, 6282–6286.

Mao, H.-K., Bell, P. M., Shaner, J. W. and Steinberg, D. J. (1978). Specific volume measurements of Cu, Mo, Pd, and Ag and calibration of the ruby R1 fluorescence pressure gauge from 0.06 to 1 Mbar. *Journal of Applied Physics*, **49**, 3276–3283.

Mao, H. K. and Hemley, R. J. (1994). Ultrahigh-pressure transitions in solid hydrogen. *Reviews of Modern Physics*, **66**, 671–692.

Mao, Z., Dorfman, S. M., Shieh, S. R., Lin, J. F., Prakapenka, V. B., et al. (2011). Equation of state of a high-pressure phase of $Gd_3Ga_5O_{12}$. *Physical Review B*, **83**, 054114-1–054114-6.

Marsh, S. P., ed. (1980). *LASL Shock Hugoniot Data*. Berkeley: University of California Press.

Mashimo, T., Chau, R., Zhang, Y., Kobayoshi, T., Sekine, T., et al. (2006). Transition to a virtually incompressible oxide phase at shock pressures of 120 GPa (1.2 Mbar): $Gd_3Ga_5O_{12}$. *Physical Review Letters*, **96**, 105504-1–105504-4.

Mattesson, T. R., Root, S., Mattsson, A. E., Shulenburger, L., Magyar, R. J. and Flicker, D. G. (2014). Validating density-functional theory simulations at high energy-density conditions with liquid krypton shock experiments to 850 GPa on Sandia's Z machine. *Physical Review B*, **90**, 184105-1–184105-10.

McMahan, A. K. (1976). *Bulletin of the American Physical Society*, **21**, 1303.

McMahon, J. M. and Ceperley, D. M. (2011). Ground-state structures of atomic metallic hydrogen. *Physical Review Letters*, **106**, 165302-1–165302-4.

McMinis, J., Clay III, R. C., Donghwa, L., and Morales, M. A. (2015). Molecular to atomic phase transition in hydrogen under pressure. *Physical Review Letters*, **114**, 105305-1–105305-4.

McQueen, R. G., Marsh, S. P., Taylor, J. W., Fritz, J. N. and Carter, W. J. (1970). The equation of state of solids from shock wave studies. In *High-Velocity Impact Phenomena*, ed. R. Kinslow. New York: Academic, pp. 293–417, 515–568.

McQueen, R. G., Hopson, J. W. and Fritz, J. N. (1982). Optical technique for determining rarefaction wave velocities at very high pressures. *Review of Scientific Instruments*, **53**, 245–250.

McWilliams, R. S., Dalton, D. A., Mahmood, M. F. and Goncharov, A. F. (2016). Optical properties of fluid hydrogen at the transition to the conducting state. *Physical Review Letters*, **116**, 255501-1–255501-6.

Medvedev, A. B. (2010). Crater formation possibly associated with an ascending thermal plume. *New Concepts in Global Tectonics Newsletter*, 56, 86–98.

Melosh, H. J. (1989). *Impact Cratering*. New York: Oxford University Press.

Mendelssohn, K. (1966). *The Quest for Absolute Zero*. New York: World University Library.

Mermin, N. D. and Ashcroft, N. W. (2006). Hans Bethe's contributions to solid-state physics. *International Journal of Modern Physics B*, **20**, 2227–2236.

Meyers, M. A. (1994). *Dynamic Behavior of Materials*. New York: Wiley.

Mikaelian, K. O. (1985). Richtmyer-Meshkov instabilities in stratified fluids. *Physical Review A*, **31**, 410–419.

Militzer, B. and Ceperley, D. M. (2000). Path integral Monte Carlo calculation of the deuterium Hugoniot. *Physical Review Letters*, **85**, 1890–1893.

Mintsev, V. B. and Fortov, V. E. (2015). Transport properties of warm dense matter behind intense shock waves. *Laser and Particle Beams*, **33**, 41–50.

Mitchell, A. C. and Nellis, W. J. (1981a). Diagnostic system of the Lawrence Livermore National Laboratory two-stage light-gas gun. *Review of Scientific Instruments*, **52**, 347–359.

Mitchell, A. C. and Nellis, W. J. (1981b). Shock compression of aluminum, copper, and tantalum. *Journal of Applied Physics*, **52**, 3363–3374.

Mitchell, A. C. and Nellis, W. J. (1982). Equation of state and electrical conductivity of water and ammonia shocked to the 100 GPa (1 Mbar) pressure range. *Journal of Chemical Physics*, **76**, 6273–6281.

Mitchell, A. C., Nellis, W. J., Moriarty, J. A., Heinle, R. A., Tipton, R. E. and Repp, G. W. (1991). Equation of state of Al, Cu, Mo, and Pb at shock pressures up to 2.4 TPa (24 Mbar). *Journal of Applied Physics*, **69**, 2981–2986.

Mochalov, M. A., Glukhodedov, V. D., Kirshanov, S. I. and Lebedeva, T. S. (2000). Electric conductivity of liquid argon, krypton and xenon under shock compression up to pressure of 90 GPa. In *Shock Compression of Condensed Matter – 1999*, eds. M. D. Furnish, L. C. Chhabildas and R. S. Hixson. New York: American Institute of Physics, pp. 983–986.

Morales, M. A., Pierleoni, C., Schwegler, E. and Ceperley, D. M. (2010). Evidence for a first-order liquid-liquid transition in high-pressure hydrogen from ab initio simulations. *Proceedings of the National Academy of Sciences*, **107**, 12799–12803.

Mott, N. F. (1934). The resistance of liquid metals. *Proceedings of the Royal Society of London. Series A*, **146**:857, 465–472.

Mott, N. F. (1936). *The Theory of the Properties of Metals and Alloys*. New York: Dover.

Mott, N. F. (1972). Conduction in non-crystalline systems IX. The minimum metallic conductivity. *Philosophical Magazine*, **26**, 1015–1026.

Motz, H. (1979). *The Physics of Laser Fusion*. Academic: London.

Nellis, W. J. (1999). Metastable solid metallic hydrogen. *Philosophical Magazine B*, **79**, 655–661.

Nellis, W. J. (2000). Making metallic hydrogen. *Scientific American*, **282**, 84–90.

Nellis, W. J. (2002a). In *High Pressure Phenomena*. ed. R. J. Hemley, G. L. Chiarotti, M. Bernasconi and L. Ulivi. Amsterdam: IOS Press.

Nellis, W. J. (2002b). Shock compression of deuterium near 100 GPa pressures. *Physical Review Letters*, **89**, 165502-1–165502-4.

Nellis, W. J. (2003). Shock-compression of a free-electron gas. *Journal of Applied Physics*, **94**, 272–275.

Nellis, W. J. (2005). High pressure effects in supercritical rare-gas fluids. In *Electronic Excitations in Liquefied Rare Gases*, eds. W. F. Schmidt and E. Illenberger. Stevenson Ranch, CA: American Scientific Publishers, pp. 29–50.

Nellis, W. J. (2006a). Dynamic compression of materials: metallization of fluid hydrogen at high pressures. *Reports on Progress in Physics*, **69**, 1479–1580.

Nellis, W. J. (2006b). Sensitivity and accuracy of Hugoniot measurements at ultrahigh pressures. In *Shock Compression of Condensed Matter – 2005*, eds. M. D. Furnish, M. Elert, T. P. Russell and C. T. White. Melville, NY: American Institute of Physics, pp. 115–118.

Nellis, W. J. (2007a). Discovery of metallic fluid hydrogen at 140 GPa and ten-fold compressed liquid density. *Review of High Pressure Science and Technology (Japan)*, **17**, 328–333.

Nellis, W. J. (2007b). Adiabat-reduced isotherms at 100 GPa pressures. *High Pressure Research*, **27**, 393–407.

Nellis, W. J. (2010). P. W. Bridgman's contributions to the foundations of shock compression of condensed matter. *Journal of Physics: Conference Series*, **215**, 012144-1–012144-4.

Nellis, W. J. (2012). Possible magnetic fields of super Earths generated by convecting, conducting oxides. *American Institute of Physics Conference Proceedings*, **1426**, 863–866.

Nellis, W. J. (2013). Wigner and Huntington: The long quest for metallic hydrogen. *High Pressure Research*, **33**, 369–376.

Nellis, W. J. (2015a). Dynamic high pressure: Why it makes metallic fluid hydrogen. *Journal of Physics and Chemistry of Solids*, **84**, 49–56.

Nellis, W. J. (2015b). The unusual magnetic fields of Uranus and Neptune. *Modern Physics Letters B*, **29**, 1430018-1–1430018-29.

Nellis, W. J. (2017). Magnetic fields of Uranus and Neptune: Metallic fluid hydrogen. In *Shock Compression of Condensed Matter (the Proceedings of the 19th Biennial APS*

Conference on Shock Compression of Condensed Matter), eds. T. G. Germann, R. Chau and T. D. Sewell.

Nellis, W. J. and Mitchell, A. C. (1980). Shock compression of liquid argon, nitrogen, and oxygen to 90 GPa (900 kbar). *Journal of Chemical Physics*, **73**, 6137–6145.

Nellis, W. J., Mitchell, A. C., Ross, M. and van Thiel, M. (1980). Shock compression of liquid methane and the principle of corresponding states. In *High Pressure Science and Technology Vol. 2*, eds. B. Vodar and Ph. Marteau. Oxford: Pergamon, pp. 1043–1047.

Nellis, W. J., Ree, F. H., van Thiel, M. and Mitchell, A. C. (1981). Shock compression of liquid carbon monoxide and methane to 90 GPa (900 kbar). *Journal of Chemical Physics*, **75**, 3055–3063.

Nellis, W. J., Mitchell, A. C., van Thiel, M., Devine, G. J., Trainor, R. J. and Brown, N. (1983). Equation-of-state data for molecular hydrogen and deuterium at shock pressures in the range 2–76 GPa (20–760 kbar). *Journal of Chemical Physics*, **79**, 1480–1486.

Nellis, W. J., Holmes, N. C., Mitchell, A. C., Trainor, R. J., Governo, G. K., Ross, M. and Young, D. A. (1984a). Shock compression of liquid helium to 56 GPa (560 kbar). *Physical Review Letters*, **53**, 1248–1251.

Nellis, W. J., Holmes, N. C., Mitchell, A. C. and van Thiel, M. (1984b). Phase transition in fluid nitrogen at high densities and temperatures. *Physical Review Letters*, **53**, 1661–1664.

Nellis, W. J., Ree, F. H., Trainor, R. J., Mitchell, A. C. and Boslough, M. B. (1984c). Equation of state and optical luminosity of benzene, polybutene, and polyethylene shocked to 210 GPa (2.1 Mbar). *Journal of Chemical Physics*, **80**, 2789–2799.

Nellis, W. J., Hamilton, D. C., Holmes, N. C., Radousky, H. B., Ree, F. H., Mitchell, A. C. and Nicol, M. (1988a). The nature of the interior of Uranus based on studies of planetary ices at high dynamic pressure. *Science*, **240**,779–781.

Nellis, W. J., Maple, M. B. and Geballe, T. H. (1988b). Synthesis of metastable superconductors by high dynamic pressure. In *SPIE Vol. 878 Multifunctional Materials*, ed. R. L. Gunshor. Bellingham: Society of Photo-Optical Instrumentation Engineers, pp. 2–9.

Nellis, W. J., Radousky, H. B., Hamilton, D. C., Mitchell, A.C., Holmes, N. C., Christianson, K. B. and van Thiel, M. (1991a). Equation-of-state, shock-temperature, and electrical-conductivity data of dense fluid nitrogen in the region of the dissociative phase transition. *Journal of Chemical Physics*, **94**, 2244–2257.

Nellis, W. J., Mitchell, A. C., Ree, F. H., Ross, M., Holmes, N. C., Trainor, R. J. and Erskine, D. J. (1991b). Equation of state of shock-compressed liquids: Carbon dioxide and air. *Journal of Chemical Physics*, **95**, 5268–5272.

Nellis, W. J., Mitchell, A. C., McCandless, P. C., Erskine, D. J. and Weir S. T. (1992). Electronic energy gap of molecular hydrogen from electrical conductivity measurements at high shock pressures. *Physical Review Letters*, **68**, 2937-1941.

Nellis, W. J., Ross, M. and Holmes, N. C. (1995). Temperature measurements of shock-compressed liquid hydrogen: Implications for the interior of Jupiter. *Science*, **269**, 1249–1252.

Nellis, W. J., Weir, S. T. and Mitchell, A. C. (1996). Metallization and electrical conductivity of hydrogen in Jupiter. *Science*, **273**, 936–938.

Nellis, W. J., Holmes, N. C., Mitchell, A. C., Hamilton, D. C. and Nicol, M. (1997). Equation of state and electrical conductivity of "synthetic Uranus," a mixture of water, ammonia, and isopropanol, at shock pressure up to 200 GPa (2 Mbar). *Journal of Chemical Physics*, **107**, 9096–9100.

Nellis, W. J., Louis, A. A. and Ashcroft, N. W. (1998). Metallization of fluid hydrogen. *Philosophical Transactions of the Royal Society*, **356**, 119–138.

Nellis, W. J., Weir, S. T. and Mitchell, A. C. (1999). Minimum metallic conductivity of fluid hydrogen at 140 GPa (1.4 Mbar). *Physical Review B*, **59**, 3434–3449.

Nellis, W. J., Hamilton, D. C. and Mitchell, A. C. (2001). Electrical conductivities of methane, benzene and polybutene shock-compressed to 60 GPa (600 kbar). *Journal of Chemical Physics*, **115**, 1015–1019.

Nellis, W. J., Mitchell, A. C. and Young, D. A. (2003). Equation-of-state measurements for aluminum, copper and tantalum. *Journal of Applied Physics*, **93**, 304–310.

Norman, G., Saitov, I., Stegailov, V. and Zhilyaev, P. (2013). Atomistic modeling and simulation of warm dense matter. *Conductivity and Reflectivity. Contributions to Plasma Physics*, **53**, 300–310.

Norman, G., Saitov, I. and Stegailov, V. (2015). Plasma-plasma and liquid-liquid first-order phase transitions. *Contributions to Plasma Physics*, **55**, 215–221.

Nuckolls, J., Wood, L., Thiessen, A and Zimmerman, G. (1972). Laser compression of matter to super-high densities: Thermonuclear (CTR) applications. *Nature*, **239**, 139–142.

Ohta, K., Ichimaru, K., Einaga, M., Kawaguchi, S., Shimizu, K., Matsuoka, T., Hirao, N. and Ohishi, Y. (2015). Phase boundary of hot dense fluid hydrogen. *Scientific Reports*, **5**, 16560-1–16560-7.

Ozaki, N., Nellis, W. J., Mashimo, T., Ramzan, M., Ahuja, R., Kaewmaraya, T., Kimura, T., Knudson, M., Miyanishi, K., Sakawa, Y., Sano, T. and Kodama, R. (2016). Dynamic compression of dense oxide ($Gd_3Ga_5O_{12}$) from 0.4 to 2.6 TPa: Universal Hugoniot of fluid metals. *Scientific Reports*, **6**, 26000-1–26000-9.

Pfaffenzeller, O. and Hohl, D. (1997). Structure and electrical conductivity in fluid high-density hydrogen. *Journal of Physics: Condensed Matter*, **9**, 11023–11034.

Piermarini, G. and Block, S. (1958). The diamond anvil pressure cell. Available at www.nvlpubs.nist.gov/nistpubs/sp958-lide/100-103.pdf

Podolak, M., Hubbard, W. B. and Stevenson, D. J. (1991). In *Uranus*, eds. J. T. Bergstrahl, E. D. Miner and M. S. Matthews. Tuscon: University of Arizona Press, pp. 29–61.

Politzer, H. D. (1973). Reliable perturbative results for strong interactions. *Physical Review Letters*, **30**, 1346–1349.

Price, M. C., Solscheid, C., Burchell, M. J., Josse, L., Adamek, N. and Cole, M. J. (2013). Survival of yeast spores in hypervelocity impact events up to velocities of 7.4 km/s. *Icarus*, 222, 263–272.

Radousky, H. B., Nellis, W. J., Ross, M., Hamilton, D. C., and Mitchell, A. C. (1986). Molecular dissociation and shock-induced cooling in fluid nitrogen at high densities and temperatures. *Physical Review Letters*, **57**, 2419–2422.

Radousky, H. B., Mitchell, A. C. and Nellis, W. J. (1990). Shock temperature measurements of planetary ices: NH_3, CH_4, and "synthetic Uranus". *Journal of Chemical Physics*, **93**, 8235–8239.

Ragan, III, C. E. (1984). Shock wave experiments at threefold compression. *Physical Review A*, **29**, 1391–1402.

Rankine, W. J. M. (1870). On the thermodynamic theory of waves of finite longitudinal disturbance. *Philosophical Transactions of the Royal Society of London*, **160**, 277–288.

Ree, F. H. (1979). Systematics of high-pressure and high-temperature behavior of hydrocarbons. *Journal of Chemical Physics*, **70**, 974–983.

Rice, M. H., McQueen, R. G. and Walsh, J. M. (1958). Compression of solids by strong shock waves. *Solid State Physics*, **6**, 1–63.

Richardson, C. F. and Ashcroft, N. W. (1997). High temperature superconductivity in metallic hydrogen: Electron-electron enhancements. *Physical Review Letters*, **78**, 18–121.

Riemann, B. (1860). Uber die Fortpflanzung ebener Luftwellen von endlicher Schwingungsweite. Abhandlungen der Gesellschaft der Wissenschaften zu Gottingen. *Mathematisch-physikalische Klasse*, **8**, 43.

Root, S., Magyar, R. J., Carpenter, J. H., Hanson, D. L. and Mattsson, T. R. (2010). Shock compression of a fifth period element: liquid xenon to 840 GPa. *Physical Review Letters*, 105, 085501-1–085501-4.

Ross, M. (1996). Insulator-metal transition of fluid molecular hydrogen. *Physical Review B*, **54**, R9589–R9591.

Ross, M. (1998). Linear-mixing model for shock-compressed liquid deuterium. *Physical Review B*, **58**, 669–677.

Ross, M., Nellis, W. and Mitchell. (1979). Shock-wave compression of liquid argon to 910 kbar. *Chemical Physics Letters*, **68**, 532–535.

Ross, M. and Ree, F. H. (1980). Repulsive forces of simple molecules and mixtures at high density and temperature. *Journal of Chemical Physics*, **73**, 6146–6152.

Ross, M., Ree, F. H. and Young, D. A. (1983). The equation of state of molecular hydrogen at very high density. *Journal of Chemical Physics*, **79**, 1487–1494.

Rossler, U. (1976). Formation and trapping of electron excitations in neon cryocrystals. In *Rare Gas Solids, Vol. I.* eds. M. L. Klein and J. A. Venables. New York: Academic, p. 545.

Rozsnyai, B. F., Albritton, J. R., Young, D. A., Sonnad, V. N. and Liberman, D. A. (2001). Theory and experiment for ultrahigh pressure shock Hugoniots. *Physics Letters A*, **291**, 226–231.

Ruoff, A. L. (1967). Linear shock-velocity-particle-velocity relationship. *Journal of Applied Physics*, **38**, 4976–4980.

Ruoff, A. L., Xia, H., Luo, H. and Vohra, Y. K. (1990). Miniaturization techniques for obtaining static pressures comparable to the pressures at the center of the earth: X-ray diffraction at 416 GPa. *Review of Scientific Instruments*, **61**, 3830–3833.

Ruzmaikin, A. A. and Starchenko, S. V. (1991). On the origin of Uranus and Neptune magnetic fields. *Icarus*, **93**, 82–87.

Sagnotti, L., Scardia, G., Giaccio, B., Liddicoat, J. C., Nomade, S., et al. (2014). Extremely rapid directional change during Matuyama-Brunhes geomagnetic polarity reversal. *Geophysical Journal International*, **199**, 1110–1124.

Saitov, I. (2016). Density functional theory for dielectric properties of warm dense matter. *Molecular Physics*, **114**, 446–452.

Schmidt, W. F. and Illenberger, E., eds. (2005). *Electronic Excitations in Liquefied Rare Gases*. Stevenson Ranch, CA: American Scientific Publishers.

Schott, G. L., Shaw, M. S. and Johnson, J. D. (1985). Shocked states from initially liquid oxygen-nitrogen systems. *Journal of Chemical Physics*, **82**, 4264–4375.

Schwager, B., Chudinovskikh, L. Gavriliuk, A. and Boehler, R. (2004). Melting curve of H_2O to 90 GPa measured in a laser-heated diamond cell. *Journal of Physics: Condensed Matter*, **16**, S1177–S1179.

Schwarzschild, B. (2003). Inertial confinement fusion driven by pulsed power yields thermonuclear neutrons. *Physics Today*, **56(7)**, 19–21.

Sekine, T. (2000). Shock wave diagnostics of sp^2 and van der Waals bonding of carbon. In Manghnani, M. H., Nellis, W. J. and Nicol, M. F., editors, *Science and Technology of High Pressure, Proceedings of AIRAPT-17*. Hyderabad: Universities Press, pp. 229–232.

Sekine, T. and Kobayashi, T. (1999). Shock-induced process during compression of graphite perpendicular to the c-axis. *Bulletin of the American Physical Society, Conference on Shock Compression of Condensed Matter*, abstract #L4.06.

Smith, R. F., Eggert, J. H., Jeanloz, R., Duffy, T. S., Braun, D. G., et al. (2014). Ramp compression of diamond to 5 terapascals. *Nature*, **511**, 330–333.

Stanley, S. and Bloxham, J. (2004). Convective-region geometry as the cause of Uranus' and Neptune's unusual magnetic fields. *Nature*, **428**, 151–153.

Stanley S. and Bloxham, J. (2006). Numerical dynamo models of Uranus' and Neptune's magnetic fields. *Icarus*, **184**, 556–572.

Starchenko, S. V. (1993). Non-axisymmetric magnetic structure generation in planets sun and galaxies. *Proceedings of the International Astronomical Union*. DOI: 10.1007/978-94-011–0772-3_48

Stevenson, D. J. (1982). Interiors of the Giant Planets. *Annual Review of Earth and Planetary Science*, **10**, 257–295.

Stevenson, D. J. (1983). Planetary magnetic fields. *Reports on Progress in Physics*, **46**, 555–620.

Stevenson, D. J. (1998). States of matter in massive planets. *Journal of Physics: Condensed Matter*, **10**, 11227–11234.

Stevenson, D. J. (2010). Planetary magnetic fields: Achievements and prospects. *Space Science Reviews*, **152**, 651–664.

Stewart, J. W. (1956). Compression of solidified gases to 20,000 kg/cm^2 at low temperature. *Journal of Physics and Chemistry of Solids*, **1**, 146–158.

Stewart, S. T., Seifter, A. and Obst, A. (2008). Shocked H2O ice: Thermal emission measurements and the criteria for phase changes during impact events. *Geophysical Research Letters*, 35, L23203-1–L23203-4.

Stokes, E. E. (1848). On a difficulty in the theory of sound. *Philosophical Magazine*, **33**, 349–356.

Strutt, J. W., Lord Rayleigh. (1883). Investigation of the character of the equilibrium of an incompressible heavy fluid of variable density. *Proceedings of the London Mathematical Society*, **14**, 170–177.

Strutt, J. W., Lord Rayleigh. (1910). Aerial plane waves of finite amplitude. *Proceedings of the Royal Society of London*, **84**, 247–284.

Subramanian, N., Goncharov, A. F., Struzhkin, V. V., Somayazulu, M. and Hemley, R. J. (2011). Bonding changes in hot fluid hydrogen at megabar pressures. *Proceedings of the National Academy of Sciences*, **108**, 6014–6019.

Tamblyn, I. and Bonev, S. A. (2010a). Structure and phase boundaries of compressed liquid hydrogen. *Physical Review Letters*, **104**, 065702.

Tamblyn, I. and Bonev, S. A. (2010b). A note on the metallization of compressed liquid hydrogen. *Journal of Chemical Physics*, **132**, 134503.

Taylor, G. I. (1950). The instability of liquid surfaces when accelerated in a direction perpendicular to their planes. I. *Proceedings of the Royal Society of London, Ser. A*, **201**, 192–196.

Thomson, W. (Lord Kelvin). (1848). On an absolute thermometric scale founded on Carnot's theory of the motive power of heat, and calculated from Regnault's observations. *Cambridge Philosophical Society, Proceedings* for June 5, 100–106.

Trainor, R. J. and Lee, Y. T. (1982). Analytic models for design of laser-generated shock-wave experiments. *Physics of Fluids*, **25**, 1898–1907.

Trunin, R. F. (1998). *Shock Compression of Condensed Materials*. Cambridge: Cambridge University Press.

Trunin, R. F., ed. (2001). *Experimental Data on Shock Compression and Adiabatic Expansion of Condensed Matter.* Sarov: Russian Federal Nuclear Center-VNIIEF.

Trunin, R. F., Boriskov, G. V., Bykov, A. I., Medvedev, A. B., Simakov, G. V. and Shuikin, A. N. (2008). Shock compression of liquid nitrogen at a pressure of 320 GPa. *JETP Letters*, **88**,189–191.

Trunin, R. F., Urlin, V. D. and Medvedev, A. B. (2010). Dynamic compression of hydrogen isotopes at megabar pressures. *Physics-Uspekhi*, **53**, 577–593.

Tubman, N. M., Liberatore, E., Pierleoni, C., Holzmann, M. and Ceperley, D. M. (2015). Molecular-atomic transition along the deuterium Hugoniot curve with coupled electron-ion Monte Carlo simulations. *Physical Review Letters*, **115**, 045301.

Urlin, V. D., Mochalov, M.A. and Mikhailova, O. L. (1997). Quasi-isentropic compression of liquid argon up to 500 GPa. *JETP*, **84**, 1145–1148.

Urtiew, P. A. (1974). Effect of shock loading on transparency of sapphire crystals. *Journal of Applied Physics*, **45**, 3490–3493.

U.S. National Aeronautics and Space Agency Voyager Program. http://www.nasa.gov/voyager.

van Thiel, M. and Alder, B. J. (1966a). Shock compression of liquid hydrogen. *Molecular Physics*, **10**, 427–435.

van Thiel, M. and Alder, B. J. (1966b). Shock compression of Argon. *Journal of Chemical Physics*, **44**, 1056–1065.

van Thiel, M., Hord, L. B., Gust, W. H., Mitchell, A. C., D'Addario, M., et al. (1974). Shock compression of deuterium to 900 Kbar. *Physics of Earth and Planetary Interiors*, **9**, 57–77.

Vargaftik, N. B. (1975). *Handbook of Physical Properties of Liquids and Gases* (Second edition). New York: Hemisphere.

Von Baeyer, H. C. (1993). *The Fermi Solution: Essays on Science*. New York: Random House.

Von Neumann, J. and Richtmyer, R. D. (1950). A method for the numerical calculation of hydrodynamic shocks. *Journal of Applied Physics*, **21**, 232–237.

Wang, Y., Ahuja, R. and Johansson, B. (2002). Reduction of shock-wave data with mean-field potential approach. *Journal of Applied Physics*, **92**, 6616–6620.

Weir, S. T., Mitchell, A. C. and Nellis, W. J. (1996a). Metallization of fluid molecular hydrogen at 140 GPa (1.4 Mbar). *Physical Review Letters*, **76**, 1860–1863.

Weir, S. T., Mitchell, A. C. and Nellis, W. J. (1996b). Electrical resistivity of single-crystal Al2O3 shock-compressed in the pressure range 91-220 GPa (0.91-2.20 Mbar).

Wigner, E. and Huntington, H. B. (1935). On the possibility of a metallic modification of hydrogen. *Journal of Chemical Physics*, **3**, 764–770.

Wilkins, M. L. (1999). *Computer Simulation of Dynamic Phenomena.* Berlin: Springer.

Yakushev, V. V., Postnov, V. I., Fortov, V. E. and Yakysheva, T. I. (2000). Electrical conductivity of water during quasi-isentropic compression to 130 GPa. *Journal of Experimental and Theoretical Physics*, **90**, 617–622.

Zaghoo, M., Salamat, A. and Silvera, I. F. (2016). Evidence of a first-order phase transition to metallic hydrogen. *Physical Review **B***, 93, 155128.

Zeldovich, Ya. B. and Raizer, Yu. P. (1966). *Physics of Shock Waves and High-Temperature Hydrodynamic Phenomena*, Volumes I and II. New York: Academic.

Zhang, D.-Y., Liu, F.-S., Hao, G.-Y. and Sun, Y.-H. (2007). Shock-induced emission from Sapphire in High-pressure phase of Rh_2O_3 (II) structure. *Chinese Physics Letters*, **24**, 2341–2344.

Zhernokhletov, M. V., Simakov, G. V., Sutulov, Yu. N. and Trunin, R. F. (1995). Isentropic compression of matter by a pulsed magnetic field. *High Temperature*, **33**, 36–272.

Zhernokhletov, M. V. (2005). *Methods for Study of Substance Properties under Intensive Dynamic Loading*. New York: Springer-Velar.

Zhou, X., Li, J., Nellis, W. J., Wang, X., Li, J., et al. (2011). Pressure-dependent Hugoniot elastic limit of $Gd_3Ga_5O_{12}$ single crystals. *Journal of Applied Physics*, **109**, 083536.

Zhou, X., Nellis, W. J., Li, J., Li, J., Zhao, W., et al. (2015). Optical emission, shock-induced opacity, temperatures, and melting of $Gd_3Ga_5O_{12}$ single crystals shock-compressed from 41 to 290 GPa. *Journal of Applied Physics*, **118**, 055903.

Zubarev, V. N. and Telegin, G. S. (1962). The impact compressibility of liquid nitrogen and solid carbon dioxide. *Soviet Physics Doklady*, **142**, 309–312.

Index